T0291529

Digital Transformations

Digital Transformations

New Tools and Methods for Mining Technological Intelligence

Tugrul U. Daim

Professor, Maseeh College of Engineering and Computer Science, Department of Engineering and Technology Management, Portland State University, OR, USA

Haydar Yalçın

Associate Professor, Division of Management Information Systems, Department of Business Administration, Faculty of Economics and Administrative Sciences, Ege University, Izmir, Turkey

Cheltenham, UK • Northampton, MA, USA

© Tugrul U. Daim and Haydar Yalçın 2022

All rights reserved. No part of this publication may be reproduced, stored in a retrieval system or transmitted in any form or by any means, electronic, mechanical or photocopying, recording, or otherwise without the prior permission of the publisher.

Published by
Edward Elgar Publishing Limited
The Lypiatts
15 Lansdown Road
Cheltenham
Glos GL50 2JA
UK

Edward Elgar Publishing, Inc.
William Pratt House
9 Dewey Court
Northampton
Massachusetts 01060
USA

A catalogue record for this book
is available from the British Library

Library of Congress Control Number: 2021949017

This book is available electronically in the **Elgar**online
Business subject collection
http://dx.doi.org/10.4337/9781789908633

Printed on elemental chlorine free (ECF)
recycled paper containing 30% Post-Consumer Waste

ISBN 978 1 78990 862 6 (cased)
ISBN 978 1 78990 863 3 (eBook)

Printed and bound in the USA

Contents

Introduction to *Digital Transformations*

Technology management has become a powerful tool with the continued increase in emerging technologies. Technology assessment and forecasting is critical for technology managers who seek to reduce the uncertainty created by emerging technologies. With mining tools and the availability of databases, technology mining (TM) has emerged as an effective technology management tool.

Prior research has focused on bibliometrics, patent analysis, and network analysis, as outlined below. This volume tries to integrate these methods to uncover further intelligence.

This book is organized according to the three methods mentioned above. The objective is to demonstrate the application of each tool through multiple cases. We chose cases using the lists published by leading associations and institutes as well as our discussions with colleagues. We recommend the use of multiple approaches to be able to see the whole picture. We believe that the cases chosen represent important technologies enabling a digital transformation of our lives and how we do business.

This book provides important knowledge for the following audiences:

- Industry and government professionals: This book provides multiple benefits for working professionals. First, it demonstrates the tools for mining technological intelligence out of publications and patents. Second, it applies the tools to emerging technologies enabling digital transformation.
- Researchers: This book introduces the researchers in the areas of technology and innovation management. Future research can easily be extended through the case applications in this book. Cases can be analyzed using all three emerging approaches, which can be integrated with traditional approaches such as growth curves or scenarios to provide a deeper analysis for each case.
- Professors: The cases in the book can be used in classes focusing on technology management. The cases would enable discussions about how to identify emerging technologies as well as the leaders in the field.
- Students: Graduate students in programs such as Technology Management, Business Administration, Science and Technology Studies and Engineering Management can use this book as a reference in their studies.

BIBLIOMETRICS

Bibliometrics is used to review publications, including journal and conference papers. Reviews of previously published research topics have been used for many years and provide vital intelligence for future research. Prior research includes reviews of many technologies using bibliometrics. Garces et al. (2017) identified leading researchers and institutes in energy-efficient advanced commercial refrigeration using bibliometrics. Behkami and Daim (2012) studied health information technologies using bibliometrics. Marzi et al. (2017) explored the evolution of the literature on product and process innovation through bibliometrics. Finally, Daim and Suntharasaj (2009) integrated bibliometrics into a Bass model to explore technology diffusion. Daim et al. (2006) conducted one of the first studies on the integration of bibliometrics with growth curves. Figures 0.1 and 0.2 provide examples for the topic of fuel cells.

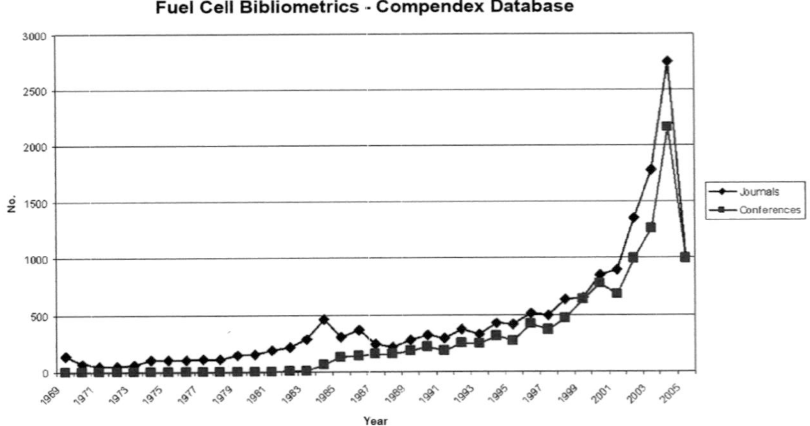

Source: Adopted from Daim et al. (2006).

Figure 0.1 *Fuel cell bibliometrics – compendex index*

There are interesting points to focus on in these types of analyses. Generally, it is misleading to look at the publications from the current year because the total is still cumulating. There are two main kinds of publications. The first are conference publications and the second are journal publications. One must

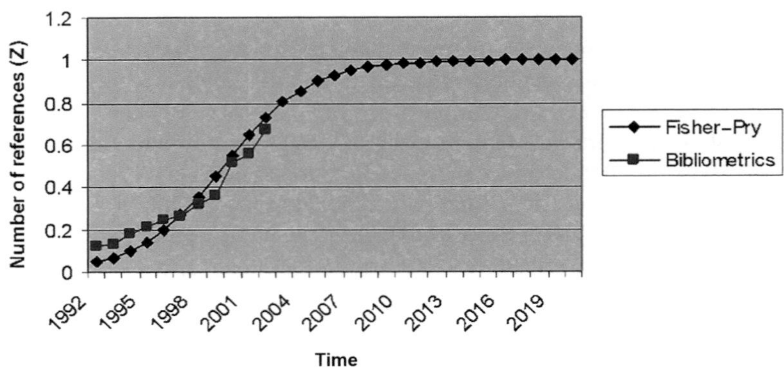

Source: Adopted from Daim et al. (2006).

Figure 0.2 Maturity model

be careful in interpreting the results. Journal publications generally take two to three years to appear in indices whereas conference publications appear much faster. Research featured in journal publications may be as dated as six to seven years old. Conferences, on the other hand, publish the most recent research. So, we need to interpret the results considering these timelines.

When we plot publications, we may not see many journal publications in an emerging area whereas conference publications in the same area may be on the rise. Similarly, we may see journal articles increasing in a mature area whereas conference publications may be going down.

Bibliometrics tends to identify activities in the academic world, as industrial actors do not tend to publish papers. Patents are a better indicator of industrial progress.

PATENTS

While bibliometrics gives us insight into academic progress, patents provide the same intelligence for industrial progress. Prior research has applied this approach to many areas. Fallah et al. (2012) leveraged patent analysis to study the movement of engineers between different companies. Other studies have used patent analysis to explore wind energy (Daim et al., 2012), smart buildings (Madani et al., 2017), televisions (Cho & Daim, 2016), and electric cars (Gibson et al., 2017).

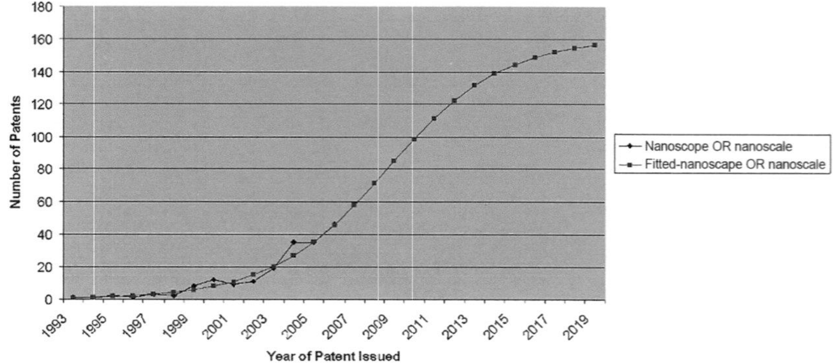

Source: Adopted from Daim et al. (2007).

Figure 0.3 Growth curves for patent data

Figure 0.3 shows how patent data can be used to fit the development to a growth curve, which is also known as an S curve.

Recent research has advanced the applications of bibliometrics, including leveraging patent data to help with decision making (Gonçalves et al., 2019) and to predict future patent citations (Madani et al., 2018).

NETWORK ANALYSIS

Network analysis provides depth to intelligence by using either the patent or bibliometrics applications. As demonstrated throughout this book, network analysis can provide metrics for any network and helps us to identify the importance of each node. Recent research has advanced the use of patent analysis alongside network analysis, as with the case of autonomous vehicles studied by Li et al. (2019).

Figures 0.4 and 0.5 provide two metrics of such an analysis applied to technology roadmapping, which is an approach used in technology management.

The data in Figures 0.4 and 0.5 were acquired from the University of Cambridge and the analysis is provided as a demonstration of the approach.

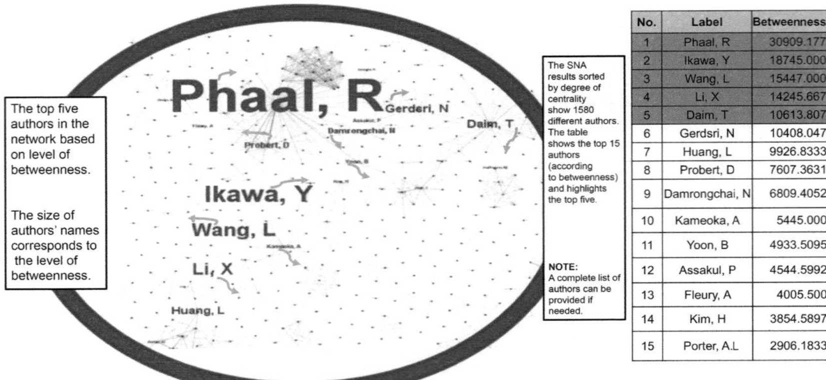
The top five authors in the network based on level of betweenness.
The size of authors' names corresponds to the level of betweenness.
Phaal, R
Gerderi, N
Daim, T
Ikawa, Y
Wang, L
Li, X
Huang, L
The SNA results sorted by degree of centrality show 1580 different authors. The table shows the top 15 authors (according to betweenness) and highlights the top five.
NOTE: A complete list of authors can be provided if needed.

No.	Label	Betweenness
1	Phaal, R	30909.177
2	Ikawa, Y	18745.000
3	Wang, L	15447.000
4	Li, X	14245.667
5	Daim, T	10613.807
6	Gerdsri, N	10408.047
7	Huang, L	9926.8333
8	Probert, D	7607.3631
9	Damrongchai, N	6809.4052
10	Kameoka, A	5445.000
11	Yoon, B	4933.5095
12	Assakul, P	4544.5992
13	Fleury, A	4005.500
14	Kim, H	3854.5897
15	Porter, A.L	2906.1833

Figure 0.4 *Authors based on degree centrality*

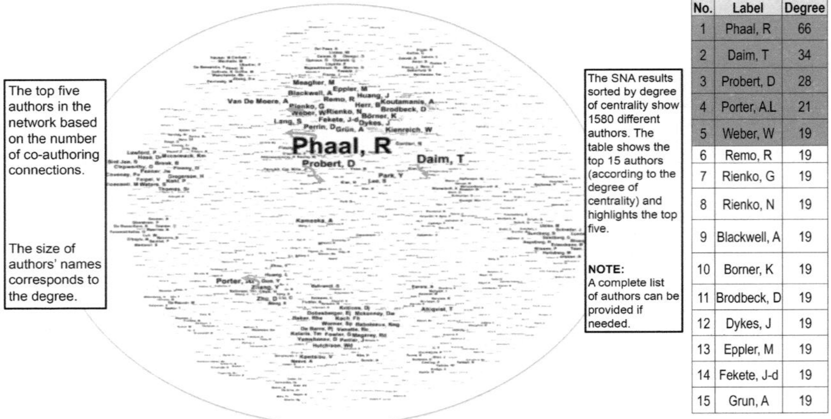
The top five authors in the network based on the number of co-authoring connections.
The size of authors' names corresponds to the degree.
Phaal, R
Probert, D
Daim, T
The SNA results sorted by degree of centrality show 1580 different authors. The table shows the top 15 authors (according to the degree of centrality) and highlights the top five.
NOTE: A complete list of authors can be provided if needed.

No.	Label	Degree
1	Phaal, R	66
2	Daim, T	34
3	Probert, D	28
4	Porter, A.L	21
5	Weber, W	19
6	Remo, R	19
7	Rienko, G	19
8	Rienko, N	19
9	Blackwell, A	19
10	Borner, K	19
11	Brodbeck, D	19
12	Dykes, J	19
13	Eppler, M	19
14	Fekete, J-d	19
15	Grun, A	19

Figure 0.5 *Authors based on betweenness centrality*

CONCLUSION

The following table lists the technology cases and corresponding methods used throughout the rest of the book.

As seen in the following table, we tried to focus on emerging technologies and their emerging applications. We were working on this book when the COVID-19 pandemic hit the world. We therefore decided to focus on it in the first chapter. The medical field was considered one of the most important application areas. In addition to the first chapter, four other chapters explore

Table 0.1 Technology case list

Bibliometrics	COVID-19
	Medical AI
	Robotic surgery
	Transgenic fish
Patent analysis	Bioprinting
	Medical 3D scanning
	Wireless power
	Drones in agriculture
Network analysis	Automated vehicles
	Electric vehicles
	Smart homes
	Space travel
Integrated analysis	Digital twin
	Supercomputing

applications of emerging technologies in the medical field: artificial intelligence, robotic surgery, bioprinting, and three-dimensional (3D) scanning. We included three chapters focusing on emerging technologies in transportation: space travel and automated and electric vehicles. Finally, the remaining four chapters explore four different fields: transgenic fish, wireless power, drones in agriculture, and smart homes. In addition to these chapters, we have included a chapter on integrated analysis in which two different technologies are examined using all the studied methods and data types. One of the technologies we examine in this section is digital twinning and the other is supercomputing.

When combined, these cases present a major technological transformation of our society. The data collected clearly show that work is underway in many institutes around the world, making us believe that the transformation is on its way.

FUTURE RESEARCH

The future is already here. Many of the analyses shown in this book can be replicated with other datasets that can be acquired from different databases. For example, Lin et al. (2019) used text mining to explore solar technology. Similarly, Twitter data, litigation data, sponsored research data, and job openings can all be used to identify a perspective on any technology.

We invite all our colleagues to expand and build upon our work and further disclose the progress of digital transformation.

1. Bibliometric-based analyses

Bibliometrics aims to make inferences about a specific discipline, technology area, or type of publication. In other words, bibliometrics, which includes the analysis of scientific communication, books, magazines, and so on, aims to reveal the basic dynamics of scientific communication by applying quantitative methods to environments. In his work that gave the method its name, Pritchard defined bibliometrics as 'applying mathematical and statistical methods to scholarly communication environments' (1969, p. 3). Bibliometric studies are among the methods that are used to reveal the relationships between the actors involved in scientific communication and that aim to reveal the levels of contribution among the actors. In this regard, all actors involved in the scientific research process, such as the author, research area, citation, journal, institution, country, and so on, are included in the scope of bibliometric studies. For each analysis unit examined, it is possible to reveal the cognitive structure for a specific discipline or research subject and determine the cooperative structure for multiple disciplines, as well as obtaining collective information at these points.

Scientometric studies were used by Eugene Garfield to measure how scientific activities are structured, how they develop, and how the actors involved in scientific communication can be measured (Moed, 2005, pp. 11–12). This became possible with the publication of citation directories such as the Science Citation Index (SCI), Social Science Citation Index (SSCI), and Arts and Humanities Citation Index (A&HCI) by the Institute of Scientific Information (ISI). Journal Citation Reports (JCR) was later published by the same institution. Within these indices, journals from various subject areas are indexed. Data used in bibliometric analyses, such as author names, article titles, institution information, data cited in articles from other articles, and so on, are indexed. In addition to these indices, scientometric studies include analysis of the bibliometric features of scientific journals. It is known that the rate of use of articles published in scientific journals decreases over time (the possibility of citation decreases). A value called the half-life is used to measure this situation, which is known as literature aging (Umut & Coştur 2007, p. 4). In general terms, it is possible to say that bibliometrics is the study of research based on authors' citing habits. However, in order to better understand the bibliometric method, it is useful to know the units it analyzes and what purposes the data produced during the measurement processes can be used for. In this chapter, information is given about the analysis units used in bibliometric methods and what kind of information these units can provide us with about TM.

Bibliometrics is understood to be a tradition dating back to the 20th century, as scientists have cited previous studies about the scientific fields they belong to (Al & Soydal, 2010; Ewing, 1966; Garfield et al., 1964). In the process of scholarly communication, citations are used to emphasize that similar research has been done before. While bibliometrics aims to reveal author–author, author–institution, institution–institution, and country–country relations through citation data for the scholarly communication process, it also specifically aims to evaluate particular journals or researchers and measure their scholarly impact (Garfield, 1979). Bibliometrics has made it possible to determine the maps and information networks for scientific collaboration networks and the institutions and authors of countries that influence technological development or innovation processes.

The bibliometric method has different names due to the format of the units subject to analysis. While studies that aim to reveal the basic dynamics of a discipline or subject area for traditional purposes are called bibliometric studies, those that aim to measure science, technology, or innovative knowledge based on innovation are called scientometric studies. Studies that determine the rate of sharing and the social reputation of a product in social networks by examining the usage information from social media are called alternative metric studies (altmetrics). Scientometrics specifically deals with the analysis of interdisciplinary relationships. It uses bibliometrics as a method. Although it is called different names according to the tools and research units used, each version of the bibliometric method aims to increase the efficiency of the information while measuring the mobility of the scientific information. For example, the method that deals with the adaptation of bibliometric studies to the Internet environment is called cybermetrics. Topics such as the activities of discussion groups outside of the Web and communication via e-mail are included in the fields of cybermetrics, which also encompasses webometrics. Webometrics is a method that enables the creation and use of Web information resource structures and technologies and that allows the quantitative study of bibliometric and informative approaches. Webometric research encompasses content analysis of webpages, web-link analysis (link analysis), Web usage analysis (log analysis), and analysis of Web technologies. Bibliometric studies that make inferences by examining various analysis units are used in various disciplines, with bibliometric laws for different disciplines and different stages of the scholarly communication process (Park & Thelwall, 2008a, 2008b).

Scientometrics involves the process of analyzing and interpreting systematically organized information. Bibliometric measures can only be used in analysis when they can produce similar results in similar studies. The structure of the bibliometric criteria, which allows the use of a standard analysis system based on power laws (such as Lotka's law, Bradford's law, Pareto's law, and so on), is the most basic feature that increases confidence in the method. This situation is in harmony with the repeatability approach, which is the basis of the philosophy of science. The ambiguity of the boundaries of research areas in science studies and the rise in the number of fields created by more than one discipline appear to be important limiting factors in determining the scope of the research. Most of the classification efforts in the literature have been carried out for the classification of journals. In other words, classification efforts have been carried out at the journal level. Waltman and Van Eck (2012), who criticized the efforts carried out at the journal level and stated that such studies should be done at the article level, developed a methodology for classifying publications at the article level through direct citations, co-citation situations, and bibliographic matching methods, which they used to obtain 10 million pieces of article information. Although the researchers attempted to classify this information at the article level, they still used the subject areas of the journals to classify the articles hierarchically.

As can be seen, the mentioned constraints have led to the development of many approaches to determining the scope of research in order to increase the reliability level of the science. Glänzel and Schubert (2003) proposed two approaches, namely a 'hierarchical' method and a 'well-structured classification method', in order to avoid classification constraints for science. It is possible to talk about a series of approaches to categorizing science – in other words, reclassifying it – in science studies. If we examine these approaches, we can see that three steps stand out. The intellectual approach is the first step in the methods of classifying. In this approach, similar to what Lewison did in his work, scientific research areas are examined by a field expert using the scientometric methodology, and each research area and sub area are subjected to an iterative evaluation and divided into categories (Lewison, 1996). In the next step (the utilitarian approach), the intention is to classify scientific journals. One of the first examples of this approach, which aimed to classify journals according to their purpose and publication policies and was based on the assumption that journals with general or special subjects in the scientific literature could be classified based on their classification, is the study carried out by Narin and others. In their study, which examined more than 900 biomedical journals, the group established a classification based on journals' research level and research subject (Narin et al., 1976; Schwartz & Hellin, 1996). In the scientometric approach, the aim is to classify publications according to their content. Bibliometric methods are used to divide publications in core journals into subfields according to their content (Daim et al., 2020; De Bruin & Moed, 1993).

The next section provides an in-depth examination of the TM method, which aims to measure science, technology, or innovative information based on innovation. With increases in capacity, the importance of estimation methods for the analysis of the available data is increasing for decision makers. Data produced by computer systems are worthless by themselves because they do not make sense when viewed from a single perspective. These data begin to make sense when processed for a specific purpose. Therefore, it is important to be able to use techniques that can process large amounts of data. Transforming these raw data into information or making them meaningful can be done with data mining. Data mining is used to obtain meaningful collections of big data and to create value (Boyack et al., 2002; Zinda, 2004). To date, it has been used to discover information in many different fields. Recently, collecting data in layers by feeding them from different systems, such as sensors, electronic devices, networks, and social media, has gained great importance for those who want to achieve meaningful results in various fields (Cai et al., 2016; Marjani et al., 2017; Russom, 2011; Zikopoulos & Eaton, 2011). This, in turn, has increased the importance of data mining and contributed to its development in different fields. In general, data mining models are classified as descriptive and predictive (Kantardzic, 2011). There are several techniques used in the studies that use these models, such as general-trend descriptive and predictive mining techniques, association rules, clustering, sequential pattern mining, and classification (Jain & Srivastava, 2013). Nowadays, many research studies are being carried out in order to obtain meaningful results by analyzing big data. These studies offer effective opportunities in the field of competition and give companies a competitive advantage over competitors (Kubina et al., 2015; Prescott, 2014). At this point, the advantage of making big data meaningful can be achieved through the approaches presented in the field of data mining. In the literature, for example, Porter and Cunningham (2004) have examined TM by dividing it into seven categories. Accordingly, the analysis of the technological field is called technology intelligence. In the method called competitive intelligence, the aim is to identify important people and their interests in technology. The identification of the technology areas that meet the needs of the users or sectors in question is described as market research. A general evaluation of a technology is undertaken with technology road maps. Technology forecasts are conducted to identify technological trends. With the technology evaluation, it is possible to see potential technological fields and developments. Finally, bibliometrics is used to provide technology foresight information about alternative technologies that may be popular in the future. Here we examine three frequently used TM methods, namely bibliometrics, patent analysis, and social network analysis (SNA), and case analyses are provided for the three different technologies demonstrating each method.

PANDEMICS, R&D, AND THE QUESTION OF WHETHER THERE IS A COLLABORATIVE RESEARCH CULTURE

Coronavirus disease 2019 (COVID-19) is an infectious respiratory disease caused by severe acute respiratory syndrome coronavirus 2 (SARS-CoV-2). The disease was discovered in the city of Wuhan in the Hubei province of China in 2019. COVID-19 has spread worldwide since its discovery, causing a global pandemic. In recent history there have been various emerging epidemics, such as AIDS, Ebola, SARS, and now COVID-19. According to Rosenberg (1989), the response by humankind to these epidemics was generally reactive and humanity therefore suffered. While the struggle against viruses is usually carried out against local epidemics and endemics through non-governmental organizations or local administrations in the regions (Lurie et al., 2020), international research collaborations have also been observed in pandemics. During the pandemic period, all data related to COVID-19 were compiled in open databases, such as the Web of Knowledge, Google Scholar, Dimensions, and Lens.org, to open the way for researchers to seek COVID-19 remedies. This situation is important in terms of giving an idea about how critical it is to analyze data with the correct methods. According to public health experts, one of the most important issues in dealing with epidemic cases is the chain of infection because breaking the chain of infection is the key point in terms of preventing spread. The faster the spread is reduced and taken under control, the sooner social recovery can be completed. The data available must be analyzed correctly and transformed into supportive information for decision makers. In this regard, it is very important for there to be holistic studies that examine the research foci together with the early species of the COVID-19 virus in order to describe the general status of COVID-19 studies and to determine the trend formed by research on this subject.

In the light of this information, we can list the questions addressed in our research as follows:

1. What are the leading institutions in pandemic research?
2. How are the research foci shaped?
3. How have collaborations been shaped for funders? How are the collaborations between countries shaped?

In order to access the dataset used in the research, a series of studies were conducted to determine the precursors of the COVID-19 virus and, as a result, the concepts and terms to be used in the query to access the data were determined.[1] In the query, bibliographic information for 49 390 publications was accessed. The accessed data were subjected to a series of preliminary studies to make them ready for analysis, and the cleared data were saved in a relational database. The time interval of the query was 1954–2020, and the average number of citations per document was calculated as 25.33. While the number of studies with a single author was 532, the number of documents per author was determined as 0.395. In terms of collaboration among authors, it was found that the collaboration index was 2.77. When the number of related publications in the literature was analyzed, it was observed that the annual growth rate was 11.09 percent for COVID-19 studies. It should be underlined that this rate represents significant growth. While this trend was observed to fit the exponential growth curve, it was also seen that the calculated R value (R^2=0.9414) was at a very high level for the explanatory ratio (Figure 1.1). In simple terms, the number of coronavirus studies has increased gradually over time. It is possible to say that the relevant literature started to peak rapidly after a certain point in terms of the number of publications. Seeing such a curve in the pandemic period we are in is an expected result because this situation indicates that countries are in search of solutions for the treatment of the virus that has caused the pandemic; it is the natural result of a period in which all possible resources are mobilized in terms of both manpower and budget.

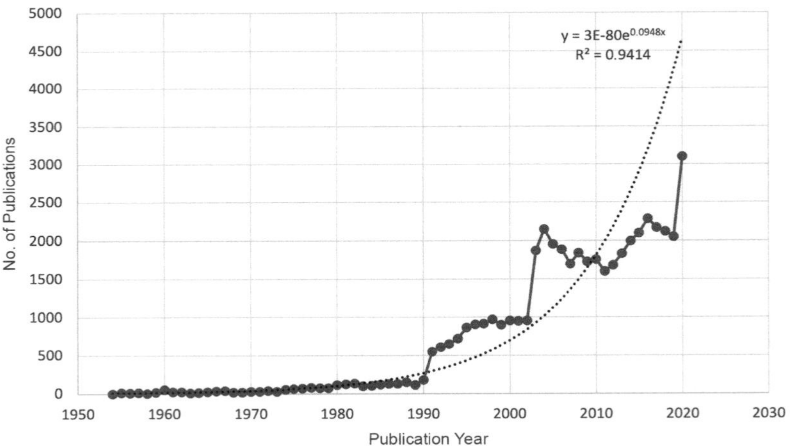

Figure 1.1 *Growth rate of coronavirus research literature*

It is possible to talk about three levels of contributions when the growing literature is examined in terms of the contributions made as a result of researchers showing such great favor to coronavirus studies: person contribution, institution contribution and country contribution.

Author Productivity

The number of publications was calculated for each author to measure author productivity. It is possible to use various methods, such as full counting and fractional counting, to determine the contribution rate in these publications. In our study, the author order of each item in the document was determined and a weighted calculation was carried out for the order. This method, also known as the dominance factor, was previously developed by Kumar and Kumar for use in measuring the performance of researchers in the oil industry (Kumar & Kumar, 2008).

Methodology

The analysis was conducted to identify the research foci that emerged in the COVID-19 research that gained momentum in the process of producing solutions for the pandemic. In this context, a co-word analysis was undertaken with the principles of co-occurrence analysis. Both the number of publications and the metrics obtained as a result of citation analysis were used to determine the core sources of the research area. On the other hand, performance indicators for each author were calculated in terms of their contribution to the field, so it was possible to identify the importance of the authors whose publications dominated the field (McCarty et al., 2013). The list created according to the mentioned criteria in terms of author contributions is presented in Table 1.1. It is worth noting that the table can be used as a very important tool in terms of showing differences in the evaluation of researchers based only on the number of publications.

Author contributions are critical to identifying the dominant actors who shape the coronavirus research area. With this information, the determination of key researchers or individuals who study technology transfer can be used to provide important information for managers at the point of strategic decision making (Schilling & Shankar, 2019). Determining who has these competencies based on the country or institution scale also provides important information. In this context, lists can be prepared using the method enabled by the author list (Table 1.1) to measure in which countries the technology followed has turned into intensive research outcomes (Guan et al., 2016). When the details of the countries and their publication performances are examined, it is possible to observe that the leading countries in COVID-19 studies are the USA, China, Germany, and England (Table 1.2).

Although determining the countries enables the determination of the regions where technological know-how is concentrated (Jacsó, 2009), the results cannot provide detailed information. Taken from this point of view, it is critical to make institutional-level determinations of technology ownership

Table 1.1 Coronavirus research author productivity

No.	Author name	h-core citation sum	All citations	All articles	h-index
1	Yuen, K.-Y.	6 311	7 988	142	51
2	Lai, M.	4 299	5 328	90	48
3	Drosten, C.	5 152	6 514	125	45
4	Yuen, K.	9 626	10 269	77	43
5	Baric, R. S.	3 620	4 885	135	41
6	Rottier, P.	4 281	4 867	69	40
7	Peiris, J.	5 995	6 323	54	39
8	Guan, Y.	7 954	8 145	49	39
9	Chan, K.	7 014	7 264	51	38
10	Enjuanes, L.	3 348	3 850	71	38
11	Weiss, S.	2 711	3 381	78	38
12	Horzinek, M.	4 065	4 510	62	38
13	Holmes, K.	4 102	4 497	62	37
14	Stohlman, S.	3 154	3 561	62	36
15	Spaan, W.	4 031	4 392	53	36
16	Perlman, S.	2 507	3 581	122	35
17	Memish, Z. A.	3 633	4 240	90	35
18	Chan, K.-H.	4 318	4 627	61	35
19	Jiang, S.	2 033	2 843	96	32
20	Snijder, E. J.	3 072	3 651	73	31

to elaborate strategic information. This approach is a tool that can be used for point-shot determinations, as well as to determine the institutions and countries that use the lists containing the main performance indicators. According to the criteria in Table 1.3, the University of Hong Kong, the University of North Carolina, Harvard University, and the Chinese Academy of Sciences are the major institutions for COVID-19 research.

In TM, after determining the main actors, there are various levels of studies that can be used when the subareas of technology are analyzed. When the literature is examined, it can be seen that frequent co-occurrence and co-word analysis are preferred for this purpose (Chen et al., 2016; Ding et al., 2001; Hu & Zhang, 2015; Topalli & Ivanaj, 2016; Wu & Leu, 2014).

Identification of Research Foci

In our study, the keywords used in defining documents within the same document were analyzed. While observing that there were five main clusters

Table 1.2 Coronavirus country productivity

No.	Country name	h-core citation sum	All citations	All articles	h-index
1	United States	210 788	1 251 096	31 797	329
2	People's Republic of China	113 422	398 651	17 140	241
3	Germany	59 038	191 513	5 663	168
4	Canada	59 181	168 459	4 888	160
5	United Kingdom	54 877	182 137	5 661	159
6	Japan	38 229	165 414	6 740	150
7	Netherlands	44 047	110 496	2 448	142
8	France	27 610	124 488	5 279	129
9	Australia	23 094	70 644	2 825	113
10	Italy	22 657	88 068	4 156	109
11	Saudi Arabia	23 351	50 015	1 851	109
12	Switzerland	24 308	54 757	1 545	103
13	Spain	15 401	54 346	2 187	95
14	Sweden	21 024	40 288	1 089	88
15	Denmark	16 673	30 958	804	87
16	Taiwan	13 643	57 437	3 117	85
17	South Korea	10 464	49 048	3 604	82
18	Belgium	11 715	25 088	930	75
19	Singapore	15 582	35 369	1 513	70
20	Finland	12 385	22 711	653	68

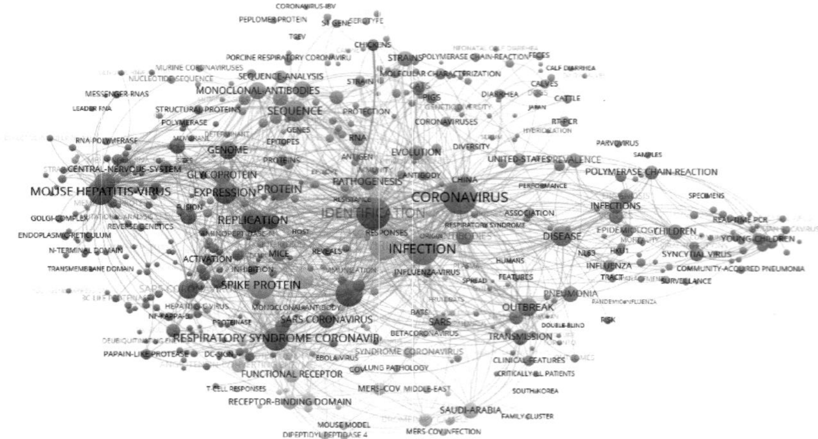

Figure 1.2 Co-word analysis of COVID-19 documents

Table 1.3 Coronavirus research institution productivity

No.	Institution name	h-core citation sum	All citations	All articles	h-index
1	University of Hong Kong	36 309	73 204	1 469	119
2	University of North Carolina	23 019	42 673	787	107
3	Harvard University	28 737	40 939	517	107
4	Chinese Academy of Sciences	14 776	31 654	1 112	82
5	University of Texas	13 068	22 699	407	81
6	University of Southern California	11 892	22 372	392	78
7	Chinese University of Hong Kong	17 954	34 254	949	77
8	Vanderbilt University	9 707	19 159	458	76
9	Centers for Disease Control and Prevention	26 060	32 541	443	75
10	Utrecht University	10 676	20 540	467	73
11	University of Iowa	8 698	19 850	530	71
12	National Institute of Allergy and Infectious Diseases	10 485	16 182	330	71
13	Scripps Research	12 739	18 597	317	70
14	University of California, San Francisco	17 794	22 490	285	70
15	Cornell University	10 124	15 987	357	67
16	Leiden University	10 854	16 698	326	67
17	University of Washington	11 621	15 747	278	66
18	University of Toronto	14 170	24 117	694	66
19	Osaka University	10 555	15 765	358	66
20	Columbia University	10 363	14 225	298	64

according to the results of the co-word analysis, there was heightened intensity in the use of the concepts for the diagnosis of the virus, as can be seen from the details of the interrelationships of the concepts. When the clusters were examined closely, it could be observed that studies were clustered as public health studies, experimental studies, descriptive studies, and vaccine studies. Figure 1.2 shows the relations of COVID-19 documents with their subfields.

In the same analysis, when we analyzed the clusters that emerged by using multiple correspondence analysis (MCA), the predominant clusters could be detected in more depth (Den et al., 2010). In order to identify and represent the foundational structures in COVID-19 documents, keywords were represented as points in low-dimensional Euclidean space as nominal categorical data (Figure 1.3).

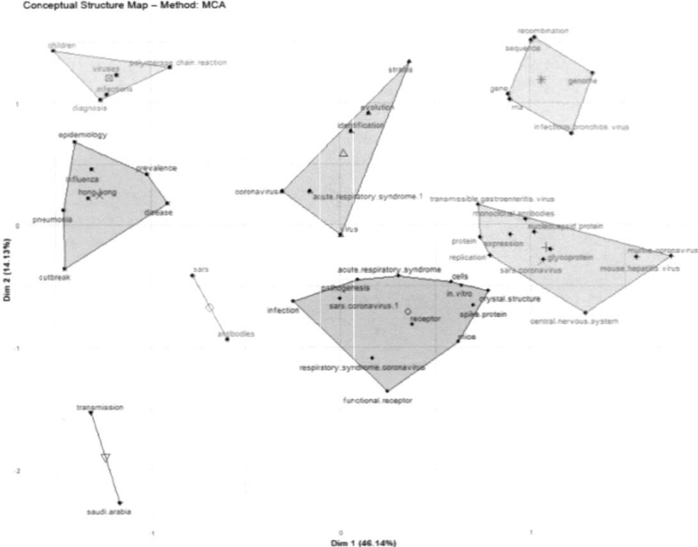

Figure 1.3 *Multiple correspondence analysis and COVID-19 document clusters*

It can be seen that the results from this section in which MCA was conducted are parallel to the results of the keyword analysis.

Collaboration Patterns in COVID-19 Research

Publications about COVID-19 should be analyzed according to subpopulation treatment effect pattern plot (STEPP) analysis principles. In this respect, it is important to identify the major actors of the research activities and to determine the research foci, as well as the collaborations of these actors with each other. For this purpose, the principles of SNA were used to determine collaborations between countries and institutions.

Taking a closer look at the collaboration between different countries, it is apparent that each country collaborates the most with researchers within its own country's institutions (Figure 1.4). This result is similar to the results of previous studies on the coronavirus literature (Yalçin & Şeker, 2020). It can be said that this situation stems from the search for the novel information required for diagnosis and treatment of COVID-19. This situation presents a parallel number to that seen in the keyword analysis. It is worth mentioning that this result is due to the requirement for specificity in the diagnosis and treatment of

Figure 1.4 Country collaborations in COVID-19 research

the virus. The search for a solution to the virus that the pandemic has caused with the internal resources of each research institution is critical in terms of both patenting and having original knowledge.

Research funding is among the important instruments used in supporting original ideas in technology, as well as in many other fields. The handling of COVID-19 studies according to the number of funds they receive can be expected to provide an opportunity to draw inferences about the originality of the studies, as well as providing important information about the cooperation patterns. In this context, the top ten studies with the most funding support in COVID-19 research are given in Table 1.4. When considered in terms of funding dynamics, the determination of international-scale studies can be undertaken more clearly. As we can see in the table, large-scale studies aiming to develop diagnosis and treatment are the ones primarily supported in terms of funders.

Conclusions

Although coronavirus research, which has turned into a race against time, has led to the maximum level of collaboration among researchers, it can still be observed that there is an introverted structure in collaboration maps. It can be seen that the necessity of following approaches where different disciplines are united and the common mind is prioritized in order to urgently find effective solutions against the common threat posed by the coronavirus is increasing. In parallel with the first detection of the virus in China, the first broadcasts were published with Chinese addresses, which can be seen as a normal result.

Table 1.4 *Research on COVID-19 with the most funding*

Number of funding sources	Study title
47	Sungnak, W., Huang, N., Bécavin, C., Berg, M., Queen, R., Litvinukova, M., Worlock, K. B., et al. (2020). SARS-CoV-2 entry factors are highly expressed in nasal epithelial cells together with innate immune genes. *Nature Medicine*, 1–7.
25	Sato, K., Misawa, N., Takeuchi, J. S., Kobayashi, T., Izumi, T., Aso, H., Nakano, Y., et al. (2018). Experimental adaptive evolution of simian immunodeficiency virus SIVcpz to pandemic human immunodeficiency virus type 1 by using a humanized mouse model. *Journal of Virology*, 92(4), e01905, 17.
20	Papa, S. M., Brundin, P., Fung, V. S., Kang, U. J., Burn, D. J., Colosimo, C., MDS-Scientific Issues Committee, et al. (2020). Impact of the COVID-19 pandemic on Parkinson's disease and movement disorders. *Mov Disord*, 6.
19	McCaw, J. M., Howard, P. F., Richmond, P. C., Nissen, M., Sloots, T., Lambert, S. B., McVernon, J., et al. (2012). Household transmission of respiratory viruses – Assessment of viral, individual and household characteristics in a population study of healthy Australian adults. *BMC Infectious Diseases*, 12(1), 345.
18	DeVincenzo, J. P. (2012). The promise, pitfalls and progress of RNA-interference-based antiviral therapy for respiratory viruses. *Antiviral Therapy*, 17(1), 213.
18	Goodrich, J. M., Quigley, K. S., Lewis, J. C., Astafiev, A. A., Slabi, E. V., Miquelle, D. G., Hornocker, M. G., et al. (2012). Serosurvey of free-ranging Amur tigers in the Russian Far East. *Journal of Wildlife Diseases*, 48(1), 186–189.
18	Wootton, S. C., Kim, D. S., Kondoh, Y., Chen, E., Lee, J. S., Song, J. W., Lancaster, L. H., et al. (2011). Viral infection in acute exacerbation of idiopathic pulmonary fibrosis. *American Journal of Respiratory and Critical Care Medicine*, 183(12), 1698–1702.
18	Papa, S. M., Brundin, P., Fung, V. S., Kang, U. J., Burn, D. J., Colosimo, C., MDS-Scientific Issues Committee, et al. (2020). Impact of the COVID-19 pandemic on Parkinson's disease and movement disorders. *Mov Disord*, 6.
17	Feng, L., Li, Z., Zhao, S., Nair, H., Lai, S., Xu, W., Yuan, Z. et al. (2014). Viral etiologies of hospitalized acute lower respiratory infection patients in China, 2009–2013. *PloS One*, 9(6).
16	Bønnelykke, K., Vissing, N. H., Sevelsted, A., Johnston, S. L., and Bisgaard, H. (2015). Association between respiratory infections in early life and later asthma is independent of virus type. *Journal of Allergy and Clinical Immunology*, 136(1), 81–86.

It is possible to say that, based on a comparison to scientific studies in other countries where the virus has been seen, the number of citations of publications from China was at a significant level in the early period, as the effort to compile the first findings required for diagnosis and treatment through the cases experienced there enabled citations of these publications for the data presented. On the other hand, it would not be wrong to say that due to the unique characteristics of countries and geographies, international cooperation in COVID-19 research cannot reach a universal scale, at least in terms of diagnosis and treatment development. When we consider the matter in terms of funding support, it can be seen that the publications that have emerged focus on examples where the effect of the COVID-19 virus on other known chronic diseases is investigated. In this regard, it is worth noting that the concept of virus-oriented cooperation is not international, but rather national, and often involves cooperation with other units within the same institution.

In this study of coronavirus studies, it can be clearly seen that the number of studies dealing with the subject in terms of social, economic, technological, environmental, and value judgments is quite limited, since research has mostly focused on the diagnosis and treatment of COVID-19. As the information obtained about COVID-19 as the main reason for the pandemic process increases, it is thought that the number of studies being carried out for this purpose will increase. Our opinion is that inter-country collaboration, in particular, will increase following the developments in diagnosis and treatment. When considered in terms of publication dynamics, it was observed that the issues addressed in coronavirus research focus more on diagnosis and treatment development. However, it is essential to handle the issue with a multi-perspective analysis because of the effects on all humanity. In this respect, we would like to draw attention to the necessity of conducting research considered also in terms of social, political, ethical, technological, economic, and environmental dimensions, as all our known business methods have changed with the coronavirus. A multidimensional or multiple perspective(s) should be considered in subsequent research in order to include the dimensions that have as-yet been studied in a restricted or ignored manner. In this context, it is necessary to increase the studies examining the socio-cultural effects of COVID-19. How the coronavirus and pandemic process affects people's attitudes toward events and situations, as well as their interests and their views, should be examined in future studies. Investigating the impact of trend changes on businesses and institutions, the structure and mobility (increase or decrease) of national and global populations, traditions, educational levels, cultural diversity, and redefinitions of standards are some of aspects that should be studied.

In the technology dimension, which is another dimension of multiple perspective analysis, what new technologies will emerge should be examined. The impact of globalization is felt intensely today, as the product life cycle is becoming shorter, increasing the demand for new products. It is clear that the concept of a return to normal does not mean a return to the old times after a period in which the effect of the pandemic on all known ways of doing business has been felt. It is clear that there is an economic change caused by the pandemic process that has affected all countries globally. Due to the fact that various sectors have been more affected by this process, the need to stop production activities and to seek solutions for these sectors will also increase. Studies on the health system, health management, people's purchasing power, and the structure of expenditures will also be needed. Another dimension comprises environmental factors, environmental protection legislation, sterilization standards, pollution, waste management and disposal, air- and water-cleaning management, energy-saving technologies, and so on. It can be said that such topics are the dimensions that should be targeted for examination in this context. The last dimension is the political dimension of the issue. The topics that can be addressed in this context include the political situation, social insurance legislation, the definition of intervention processes and their connection to regulatory texts, market regulations, trade agreements, restrictions, taxes, and the clarity of laws. In order to determine good-practice examples and mismanagement, comparative data related to practices in the world and their use in strategy formulation could be obtained and analyzed for the necessary infrastructure for conducting proactive process management.

FROM DEEP LEARNING TO SUPERCHARGED PREDICTION: ARTIFICIAL INTELLIGENCE RESEARCH IN MEDICAL SCIENCES

The aim of this section is to describe the social and intellectual structure of artificial intelligence (AI) research in scientific papers. To do so, data were gathered from the Web of Science (WoS) and bibliometric and scientometric methods were used to analyze the data. Bibliometrics was used in the creation of visuals, and collaborations in the contexts of authors, countries, and institutions were examined. H-index citation analysis was used to reveal the intellectual structure of the areas. In this context, our research objectives were to stimulate theory and understanding about AI studies in the medical sciences and to investigate who the most influential authors in the field are. The most productive authors, the most efficient institutions, and the most productive countries were visualized in this context. On the other hand, to determine the pioneers of the field, citation analysis was used along with the number of documents, and the metrics where the name rankings were taken into consideration for the documents were compared with the authors' productivities and efficiencies using the metrics related to authorship and author rankings.

Artificial Intelligence

We are passing through a technological moment in which more data bring less privacy and security and higher speed brings less accuracy. It is evident that, as automation increases, control decreases, and this directly affects quality, but it also has a direct impact on business conduct and strategies. The analysis made up to this point showed us once again that it is still difficult to talk about advanced research in AI applications. A significant number of studies are still focused on forecasting and classification. In this respect, it can be observed that intensive efforts have been made to identify and make meaning from data that has been obtained. The advances in forecasting activities, which are at the center of the decision-making process as a result of AI applications, will be used extensively in many business lines, especially in technology and engineering management (TEM). In this respect, the ways in which AI applications are handled should be examined in depth according to the criticality of the sectors. The medical sciences (MS) are also among the disciplines where AI applications are handled intensively. AI is defined as the ability of a computer or a machine under computer control to evaluate, make decisions, and implement various activities like human intelligence does (Copeland, 2019). AI, which was first conceived in 'A Logical Calculus of Ideas Immanent in Nervous Activity' (McCulloch & Pitts, 1943), reached a turning point in 1950 when Alan Turing shared his thoughts on the possibility of creating thinking machines (Harnad, 2006). The term was first used by McCarthy et al. (1955) in 1955 as a proposal for the Dartmouth Conference in 1956. Machine learning (ML) refers to computer algorithms that can model a determined question according to the data obtained in the environment in which the problem is determined. ML and deep learning (DL), which is a further step, are considered sub-branches of AI. ML is a science that makes inferences from existing data with statistics and complex math-based algorithms.

AI studies are based on three different methods: unsupervised learning based on predicting an unknown structure using unmarked data, supervised learning based on classification and regression algorithms, and reinforcement learning based on behavioral psychology-based feedback assessments. ML and AI affect physicians, hospitals, and all other health-related areas. Minor changes in health status can sometimes be lost in seas of data of magnitudes that people cannot effectively observe. When they are handled by AI applications such as DL, they become lifesaving indicators. Technology giants such as IBM, Google, and Amazon are investing in many areas of the health sector, particularly in the diagnosis of diseases. IBM Watson is an example of these technologies that aim to diagnose diseases with AI. IBM Watson aims to achieve meaningful results and outputs, with the help of techniques such as ML and DL, from the complex data that are available, as in all other AI applications. Thus, IBM Watson provides the desired results from complex data and makes meaningful connections between diseases or findings. In other words, with the help of these connections, it can easily provide predictions of diseases that real physicians may miss. In this way, IBM Watson provides physicians with possible predictions of their patients' illnesses, almost becoming a digital physician and taking on a remarkable task in diagnosing diseases. This task can be defined as an innovative solution, especially with regard to chronic diseases, such as cancer, for which diagnosis and treatment are very important (IBM, 2019).

Previous Work

Studies on AI have an important place in the bibliometrics literature. Although there are numerous studies using various methods to better understand the technology, many of them are based on bibliometrics. While the desire to develop a tool that can be put forward as a decision support system can be observed in almost all the research, it should be stated that the efforts are shaped by sectors. For example, while there is an effort to develop a method based on natural language processing in transportation engineering (de Stefano et al., 2016), neural networks are being developed in an application for determining bankruptcy in banking systems (do Prado et al., 2016). Regardless of the sector, all these studies aim to understand and recognize new technologies (Chen et al., 2018; Cox et al., 2019; Eom, 1995, 1996; Fiala & Willett, 2015; Gupta & Dhawan, 2018; Rio-Belver et al., 2018). Using a literature-based analysis, it is possible to see AI applications most commonly in studies that aim to identify methods and tools for diagnosing diseases. For example, the data contained in the reports produced by the Japan Council for Quality Health Care were used for this purpose (Akiyama et al., 2012). Furthermore, plans for personalized health services and the development of personalized treatment procedures have been advanced by combining big data with AI research (Hueso et al., 2018). Researchers seeking to demonstrate the potential of AI in health sciences examined hundreds of thousands of animals treated in veterinary hospitals throughout the US in order to predict acute renal failure. As a result of this, it was reported that the AI system accurately predicted 90 percent of the kidneys that would lose their function 48 hours in advance (Tomasev et al., 2019). In this regard, in terms of the way it is handled in MS, AI research is concentrated in the context of diagnosis. The emergence of AI research in MS, which comprises research in many fields, notably the diagnosis of diseases, prominent researchers, institutions, and so on, and the in-depth analysis of components such as project funding for AI applications in MS, will provide access to important outputs in the context of TEM. For this purpose, a case study of TEM applications was undertaken using the bibliographic data collected from the publications indexed in WoS. In this part of the study, the ways in which AI research is handled in MS as an emerging technology area are examined.

Data and Method

An online search query was made in WoS to access the data of the publications related to AI in the MS. Bibliometrics and scientometrics were used in the analysis. Bibliometrics involves the analysis of scholarly communication environments according to mathematical and statistical laws (Pritchard, 1969). Bibliometrics is a very widely used method; for example, it has been used in technology determination (Behkami & Daim, 2012; Cowan et al., 2009; Daim et al., 2006), technology road mapping (Daim, 1997a, 1997b; Daim et al., 2012; Daim et al., 2012; Daim et al., 2015), TM (Rio-Belver & Cilleruelo, 2010; Vicente-Gomila et al., 2017), and science technology policy determination (Liang & Li, 2010; Olvera et al., 2018). With this method it is possible to describe the bibliometric profile of any journal (Al et al., 2010; Yalçin, 2010), as well as the intellectual structure of any research discipline (Ustundag et al., 2016; Yalçin & Yayla, 2016). The method has been accepted on the basis of these properties and has been used for the description of scientific structures in almost all areas of study. We used several tools to prepare the data for analysis. The R programming language was used for the cleaning, compiling, and analysis of data, and VOSviewer (Van Eck & Waltman, 2009), ggplot2, and Hive libraries were used for the creation of network graphics.

Definition of research questions

To stimulate theory and understanding about AI in MS studies, we tried to visualize both the social and intellectual structures of AI research. Therefore, we defined two research questions.

Q1. What are the most discussed topics regarding AI in MS studies?
We used co-word analysis, in which the topics frequently discussed together are visualized according to co-word analysis values. Co-word analysis can be defined as a type of analysis that examines the frequency with which concepts are used together. It is applied at various levels (document titles, document abstracts, or keywords). In our study, the keywords used in the definitions of documents were examined, and network analysis was used for the relations of the concepts with each other. In this way, important concepts related to the field were determined.

Q2. Who is the most productive author on AI in MS studies?
In the analysis of the authors' performance, in addition to showing the numbers of their publications, the authors are presented in lists that allow comparisons according to the numbers of citations and the h-index values.

Country productivity
Countries were dealt with in terms of publication numbers, h-index values, h-core citation sums, and all the citations and article numbers they had.

Institution productivity
Universities were dealt with in terms of publication numbers, h-index values, h-core citation sums, and all the citations and article numbers they had.

Descriptive Statistics about Data: Language, Document Types, and Numbers

The descriptive statistics related to the dataset discussed in the analysis can be summarized as follows: between 1978 and 2018, there were 15 581 documents in total. While the number of citations per document was 11.22, the collaboration index for the authors was 2.36. The number of documents cited per document was 3.52.

Productivity Measurements and TM

Productivity measurements are often used for expert identification and core team-building purposes. They are used as a solution to the need to determine the fundamental dynamics in the field, which has become an important requirement for digital transformation and is therefore one of the most fundamental challenges of today. In this context, it can be observed that bibliometrics and TM are the most frequently used methods in the literature for productivity measurement, which can be defined as the determination of productivity potential. In our study, productivity measurement was evaluated at the researcher, institution, and country levels. As the leading activities in R&D studies, the productivity analysis TM activities of these three levels are very important.

Author productivity
In order to identify the most productive authors or researchers in the field, the number of publications was considered as an indicator. However, it is also known that analyzing the number of documents alone is misleading (Berker, 2018). Consequently, previous studies have recommended analyzing citation numbers, journals impact factors, and h-index parameters (Egghe & Rousseau, 2006; Garfield, 2006). Therefore, in order to measure author productivity, citation numbers and h-index values are given together with the number of documents (Table 1.5). When the table is examined, the effect of this normalization can be clearly seen. For example, Wang, Y., is first in terms of the number of

publications with 121, while for the number of citations Ayache, N., is in first place, despite only having 38 articles published.

Table 1.5 Author productivity in medical artificial intelligence research

h-index	Author name	h-core citation sum	All citations	All articles
23	Ayache, N.	2 216	2 297	38
15	Shahar, Y.	703	776	32
14	Shen, D.	463	546	35
14	Zhang, S.	717	798	57
14	Wang, Y.	887	1 131	121
13	Wang, J.	513	602	75
13	Pennec, X.	993	1 028	22
13	Xu, Z.	834	864	28
13	Udupa, J.	675	685	15
13	Li, X.	559	720	78
12	Lehoang, S.	358	360	14
12	Ogiela, M.	294	366	48
12	Chen, Y.	425	484	60
12	Gao, X.	533	613	29
12	Rueckert, D.	806	883	29
11	Li, S.	763	816	41
11	Rohr, K.	499	502	13
11	Lavrac, N.	482	508	21
11	Zhang, J.	321	444	82
11	Miksch, S.	634	654	22

Institution productivity

Examining the productivity of institutions, a table like the one for author productivity emerges (Table 1.6). It can be seen that the documents from Harvard University had a higher impact than the documents from other institutions. In terms of institutional impact, it can be observed that the first three US-based institutions dominate AI studies in MS, followed by the representatives of France and China. Inria, the French National Research Institute for the Digital Sciences, which is one of the fund providers in Europe, takes its place as number four.

Table 1.6 *Institutional productivity in medical artificial intelligence research*

h-index	Institution name	h-core citation sum	All citations	All articles
29	Harvard University	2 035	2 464	76
29	University of North Carolina	2 415	3 194	121
27	Massachusetts Institute of Technology	1 676	2 294	102
26	Inria	2 977	3 169	55
24	Chinese Academy of Sciences	1 564	2 092	166
23	Stanford University	1 480	1 875	81
23	University of Pennsylvania	1 873	2 265	75
22	University of Granada	1 275	1 566	93
21	University of California, Los Angeles	1 962	2 147	65
21	Georgia Institute of Technology	1 540	1 738	70
21	National University of Singapore	778	1 011	77
20	University of Oxford	4 749	5 130	78
19	Yale University	3 224	3 340	43
19	University of Pittsburgh	948	1 174	72
19	Carnegie Mellon University	1 385	1 587	91
19	University of Manchester	1 037	1 205	58
18	University of Florida	3 063	3 388	80
18	University College London	1 072	1 341	83
18	Xidian University	1 043	1 194	53
18	Selçuk University	880	1 029	40

Country productivity

In terms of country productivity, the US is in first place. In terms of country productivity according to the list of domains, the US, China, France, and the United Kingdom are at the top of the list. While Japan has a relatively good number of publications at 498, it is still below Turkey on the list. In terms of country contributions and efficiency, the countries that are concentrated on AI research with special programs are Taiwan, Japan, Greece, and Poland. They are also at the top of the list in terms of efficiency (Table 1.7).

The presence of EU member states among the countries included in the list is also remarkable. Six countries representing the EU conducted ERA-NET joint projects, which had significant impacts on creating infrastructure; Turkey has come forward to highlight this with benefit performance indicators.

Table 1.7 Country productivity in artificial intelligence research

h-index	Country name	h-core citation sum	All citations	All articles
126	United States	46 404	104 841	5 053
68	China	8 508	26 367	3 536
63	France	9 371	19 481	1 326
63	United Kingdom	13 861	25 133	1 516
53	Netherlands	6 360	10 427	527
51	Germany	5 970	14 325	1 251
50	Spain	5 786	14 167	1 441
48	Canada	7 735	13 308	811
44	Italy	3 463	8 307	945
43	Taiwan	3 384	8 407	930
41	India	4 747	11 506	2 238
41	Turkey	3 338	6 076	498
40	Australia	4 924	8 474	609
39	Japan	2 810	6 515	1 381
38	Greece	2 945	4 824	394
32	Singapore	1 845	3 137	288
32	Switzerland	3 439	5 035	299
32	Brazil	2 021	3 975	504
30	Poland	1 386	4 036	785
30	Austria	2 014	3 716	328

Source Dynamics (Journals, Proceedings, and so on)

Knowing the lists of researchers from core journals who conduct research in any field of scientific activity is important in determining the relevant sources to look for when there is a need for refined information about a field. In this respect, the calculation of the indicators for scientific journals provides good guidance in terms of both the numbers of publications and the impact they create in the field. For this purpose, we conducted a series of analysis of journals. The journals in which AI MS research is frequently published are given in Table 1.8. It is noteworthy that the relevant studies are mostly published in journals related to computer science and related disciplines. Although the journals in which AI research is published show a trend that converges toward computer science, a structure that requires researchers of both science domains stands out in the field of applications of AI in the MS field. AI technology converges with every discipline in which it finds an application. By identifying these journals where AI research relating to health sciences is frequently published, a core resource list can be produced to follow current discussions about AI. Furthermore, it is a tool that can be used as the main source for the monitoring of research and development along the axis of technology management and digital transformation and for the determination of the movements of the predecessor and successor technologies.

Table 1.8 Top journals on artificial intelligence research in the medical sciences

h-index	Journal name	h-core citation sum	All citations	All articles
68	*Medical Image Analysis*	14 145	19 880	341
57	*Artificial Intelligence in Medicine*	6 186	14 921	571
50	*Expert Systems with Applications*	3 956	10 185	491
47	*IEEE Transactions on Image Processing*	6 907	9 062	170
43	*IEEE Transactions on Pattern Analysis and Machine Intelligence*	7 902	8 789	96
34	*Computer Vision and Image Understanding*	3 640	4 463	98
33	*Pattern Recognition*	7 993	9 486	153
32	*Applied Soft Computing*	2 009	3 182	144
30	*Pattern Recognition Letters*	9 362	10 432	141
30	*International Journal of Computer Vision*	7 481	7 907	67
29	*Decision Support Systems*	1 512	2 329	98
28	*Image and Vision Computing*	2 301	2 981	98
27	*Neurocomputing*	1 556	3 009	192
23	*IEEE Transactions on Knowledge and Data Engineering*	1 852	2 258	66
22	*Knowledge-based Systems*	1 316	1 998	98
20	*Information Fusion*	2 032	2 132	31
20	*Neural Networks*	1 597	1 716	37
18	*Engineering Applications of Artificial Intelligence*	658	948	58
18	*Journal of Mathematical Imaging and Vision*	1 528	1 824	70
18	*IEEE Transactions on Fuzzy Systems*	875	943	27

Keyword Co-occurrence Analysis

The analysis of collaborations is one of the methods that can be used to identify individuals, institutions, and organizations engaged in joint studies in any scientific discipline or subject area. With this analysis, also called co-word analysis, it is possible to access data that support technology and engineering management. In our study, we used Keyword-Plus co-word analysis to determine the themes that AI research focuses on in the health sciences. In this way, it was possible to identify the topics that are frequently discussed. As a result of the co-word analysis, it was observed that AI applications in medical sciences are dealt with for diagnostic purposes, while software, hardware, and method research aim at assisting diagnosis and the analysis of images obtained in medical imaging technologies is focused on prediction (Figure 1.5). It can clearly be seen that the topic domain, where the neural networks, prediction, selection, classification, and re-presentation topics come to the foreground, is associated with almost every cluster.

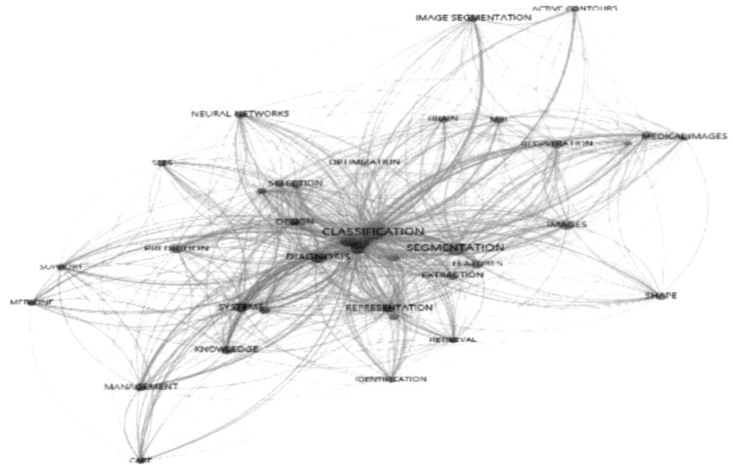

Figure 1.5 *Co-word analysis of artificial intelligence research*

Examining Table 1.9, the frequency of use of method-based concepts is remarkable. It is possible to talk about the search for a method in the research on AI in medical sciences.

Institutional Collaborations

If we examine the institutions carrying out AI research in MS and the business intelligence they have produced, it can be seen that AI technology is considered as an emerging technology area in a more capped ecosystem; in other words, institutions prefer to develop collaboration within their departments (Figure 1.6). In terms of collaboration, the tendency of research institutions to set up partnerships with the subunits within their own institutions can be attributed to the high commercial potential of the new technologies. In order

Table 1.9 Network parameters of Keyword-Plus analysis

ID	Label	Cluster	Weight (links)	Weight (total link strength)
1	Classification	4	379	2 088
2	Algorithm	1	334	1 394
3	Segmentation	1	305	1 333
4	System	2	330	1 275
5	Prediction	4	203	740
6	Diagnosis	4	243	999
7	Features	5	192	503
8	Neural networks	4	169	415
9	Model	1	365	1 570
10	Images	1	199	647
11	Selection	4	187	510
12	Algorithms	1	277	815
13	Recognition	3	207	506
14	Medical images	1	186	499
15	Neural network	4	147	316
16	Image segmentation	1	138	405
17	Support vector machines	4	125	237
18	Registration	1	189	562
19	Feature selection	4	134	323
20	Information	2	261	712

to better understand this structure of collaboration, which we can define as introverted, it is useful to discuss in depth methods such as technological knowledge redundancy.

Although it can be seen that universities based in the US are at the center of the cooperation map in terms of network values, it is possible to say that institutions with Chinese addresses also play an important role in the cooperation map. It can be observed that the Chinese Academy of Sciences and the Hong Kong University of China occupy important positions in the network according to their centrality values. It can be seen that the research is concentrated at the forecast development point. On the other hand, it is noteworthy that, when it comes to collaborations between researchers, institutions set up collaborations within their own departments. It should be noted that, as an emerging technology, such cooperative partnerships in AI research are natural. These closed patterns in terms of collaborations limit the desired level of development in AI research. In this context, it can be said that it has a destructive effect on the known business models. Among the main objectives of TM are

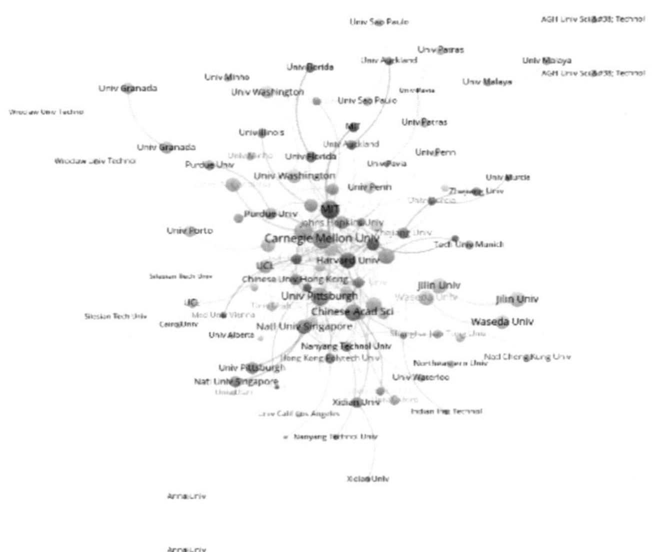

Figure 1.6 Institutional collaborations on artificial intelligence research

the identification of the methods used and the datasets applied in predecessor and successor technologies. The literature-based discovery method is one of the leading methods used for this purpose. When bibliometric methods are applied, it is possible to make inferences about vital points such as the determination of the dominant actors and of the concentration of scientific domains of a related technology. Predictions can be made about when the studies related to the technology will reach a saturation point using the growth curves for the related applications. The determination of the sub-technology domains to be invested in for the technologies that have passed from the basic research level to the product research level is among the issues examined in this context.

Discussion

AI, being clear, understandable, and practically applicable, is basically aimed at simplifying applications, and it will affect many areas of daily life. Prediction aims to complete missing data. AI is of great importance in completing such missing data. The most basic function here is that the data that are owned, but that cannot be stored in a meaningful way, are drawn from very large datasets, made meaningful, and presented to the decision maker. For this purpose, data analysis methods should be utilized in order to use the available data correctly in decision-making processes. In this way, it is possible to acquire the information necessary for companies to manage the digital transformation process. In studies where disease diagnosis and pattern recognition using medical images are a priority, the need for processes such as classification and categorization can be clearly seen. In this respect, it can be said that AI research in the medical sciences is still in its early stages. AI research in the medical sciences seems to be focused on prediction methods, as in AI research in other sectors. For AI, which has great potential in terms of technology management and digital transformation, policies that focus on the determination of hardware, software, and human resources competencies should be determined in order to identify the leading and successive technologies.

ROBOTIC TRANSFORMATION OF SURGERY: A BIBLIOMETRIC ANALYSIS OF ROBOTICS RESEARCH IN MEDICAL SCIENCES

The application of robotics in aiding clinical surgery and threading has become increasingly common in recent years, with technological advancements and the increasing interest of the research and medical sciences. This study set out to provide a comprehensive picture of the global trends and developments in robotics applications relating to the medical sciences, identifying research gaps and suggesting future directions for research and technology management. We used bibliometrics as a method. Data were extracted from the WoS database.

Introduction

Since Leonardo da Vinci's mechanical knight, designed in 1495 (Yates et al., 2011), people have sought to get machine assistance in the work they do. The dynamics of the industrial revolution have radically changed the known modes of production. In addition to these developments in mechanization, the scientific developments that accelerated in the second half of the 19th century became the basis of present day robotics research. On the other hand, the process of digitalization in the last few decades has led to many technologies, including AI, becoming solutions that are part of everyday life. At this point, robotic technologies have been put into practice in routine tasks, tasks that require intense power or that must be performed by people in extreme conditions. In other words, robots have started to perform tasks that are difficult for people to perform. While this process mostly emerges in the complex mass production lines of the automotive industry or in heavy industry, surgical robots, which are the reflection of robotic technologies to the medical sciences, also play a role as solutions that significantly increase the success rate of treatments. It is obvious that the robotics industry, which brings together disciplines such as software, engineering, and design, will contribute to military, medical, and wearable technologies, as well as in all areas of daily life (Burmaoglu et al., 2018).

Robotics studies are among the fastest developing fields of technology research. The scientific knowledge that has emerged as a result of both basic and applied research has led to tremendous development and growth in the field of robotics. Naturally, the development of these technologies and the transformations they have created in the areas they have spread to have begun to be of interest to technology and engineering management. In this respect, both the international publications in the literature and the ideas registered by patent organizations as commercial ideas have been the subject of a number of studies. When the literature is examined, it can be seen that there are many studies about robotic technologies. Although a significant majority of these studies are product research, it is possible to note that there are research foci for almost every sector. In our research, we focused on the robotic research in the medical sciences and made inferences about the intellectual, social, and theoretical structures of robotics research in the medical sector. The data obtained allowed the determination of the critical concentration in the field in the context of technology management, the identification of the dominant actors of the field (such as researchers, institutions, and so on), and the determination of the main concentrations of technology domains. In this respect, it is thought that the data obtained in this study will provide a guide for TM studies to be used in technology management.

Previous Work

Bibliometrics has been used in studies examining the development of surgical robotics studies (Shen et al., 2019). Various data visualization techniques are used in those studies in which word analysis is preferred. In another study, the use of robotic technologies in the field of urology was examined, and the countries that contributed in terms of publication production were identified (Jackson & Patel, 2019). Fan et al. (2016) conducted the largest of the studies in their work that examined studies focusing on robotic surgery between 1994 and 2015. In the research, the growth rate of the scientific studies produced in the last decade was found to be 572.87 percent, while in the address analysis it was found that the US dominates the research in the field. In another study conducted in the field of nursing, it was pointed out that, in contrast to the definition of robotic technologies, there is a more complex process in daily life than in applications (Carter-Templeton et al., 2018). When the studies that were analyzed from the data are considered in terms of their scope, it can be noted that they are limited to certain areas of robotic technologies. In this context, we examined robotic technologies in the medical sciences from cognitive, theoretical, and social perspectives.

Materials and Method

The WoS database was used to examine the ways in which robotic technologies are handled in medical sciences. Since we aimed to analyze the scientific landscape, bibliographic data from international scientific publications were analyzed. WoS records, which are accepted as an internationally authoritative source, were used for the provision of such bibliographic publications. In this way, we aimed to make inferences about the basic research trends in robotics.

Search strategy

In this study, we searched all databases of WoS for the years from 1900–2018 in order to acquire access to the bibliographic information of documents published in the medical sciences on robotic technologies. It was found that the number of studies on robotics technologies is increasing, so much so that in the WoS, which we used to access the dataset, a separate subject area for the robotics field is now defined. Between 1983 and 2018, there were 131 057 publications in the field of robotic technologies. When these publications are examined by sub area, it can be seen that approximately 48 percent of the studies are part of subject areas related to computer science.

Data extraction

The obtained data were downloaded in text format. Then, they were recorded in a relational database and prepared for analysis. Basic bibliometric indicators and power laws (such as Lotka's law, Bradford's law, and so on) were used in the analysis of the data. In the visualization of the data, social network theory and analysis methods were used. The R programming language and Excel software were used in the analysis. We then used the Perl language and related libraries to develop our own organized maps of ML methods.

Definition of research questions

We investigated robotics in MS studies. We tried to visualize both the social and intellectual structures of robotics research in MS studies. Therefore, we defined two research questions.

Q1. *What are the most discussed topics regarding robotics in MS studies?*

We used co-word analysis, in which the topics discussed frequently together are visualized according to co-word analysis values. The determination of the research domains revealed by the frequently used terms was achieved.

When these domains were evaluated collectively, it was possible to identify the emerging themes. In this respect, interpretations were made regarding the direction of research trends occurring in the MS domain.

Q2. Who is the most productive author in robotics research in MS studies?

We needed to determine the most productive authors in the field. At the same time, the number of publications per author was normalized by the authors' name rankings and the number of authors who contributed to an individual article. Identifying experts in robotics technology as an emerging technology is also important in general. However, in order to determine the expertise in medical science robotics research, which is the subject of this section, it was necessary to identify the researchers who received support or who cooperated in the development of medical science robotic technology as an emerging technology, starting from the early period and especially in developing countries. For this reason, evaluations were made with the metrics of the researchers engaged in research activities in this technology domain, depending on the number of publications.

Which country?

While this analysis provided us with strategic information regarding the planning of regional, national and international cooperation, it also highlighted important data regarding the determination of clusters formed depending on research foci. Countries were dealt with in terms of publication numbers, h-index values, h-core citation sums, and all the citations and article numbers they had.

Which university?

This analysis was important in that it enabled the identification of universities, institutions, and organizations operating in the relevant technology domain, in terms of allowing the strategic information obtained in the previous analysis to be customized at the institution and department scales. Universities were dealt with in terms of publication numbers, h-index values, h-core citation sums, and all the citations and article numbers they had (see Table 1.10).

Results

Author productivity

Although the number of documents indicated that the author productivity was not high, it was found that the studies were frequently used in terms of the number of citations they received. Although this situation draws a picture open to improvement in terms of the number of publications, it shows that the authors who were analyzed are used and cited as the main researchers of the field. In this respect, having stated that robotics research in the medical sciences is still in its early period, it can be added that the number of experts in this field is relatively low. We have already mentioned that robotic technologies are the subject of applications in many subareas as emerging technologies. The outstanding authors here had parallel values in terms of the numbers of publications and the h-index values. Another interesting result is that 3393 out of the total of 19 846 authors had not received any citations at the time of analysis (Table 1.10).

Table 1.10 *Author productivity for robotics research in the medical sciences*

h-index	Author name	h-core citation sum	All citations	All articles
12	Autorino, R.	504	513	13
12	Kaouk, J. H.	459	468	13
10	Menciassi, A.	495	543	25
10	Fichtinger, G.	442	442	10
10	Webster, R. J., III	378	409	16
10	Dario, P.	479	522	21
10	Oleynikov, D.	254	278	19
10	Dasgupta, P.	519	543	16
9	Okamura, A. M.	591	606	14
9	Laydner, H.	311	320	10
9	Stein, R. J.	323	332	10
8	Valdastri, P.	281	312	16
8	Gill, I. S.	420	427	10
8	Castle, E. P.	466	466	8
8	Kazanzides, P.	287	318	16
8	Althoefer, K.	282	295	19
8	Menon, M.	399	401	9
8	Dupont, P. E.	268	281	10
8	de Mathelin, M.	326	329	10
7	Thomas, R.	478	478	7

Country productivity

In terms of country productivity, the US was the leading country, followed by the EU member states (UK, Germany, Italy, and France). China, which was higher than Canada in terms of the number of publications, found itself in ninth place in the list in terms of h-index values. Considering the recent advances of

China in many areas, it is possible to say that it is behind in robotic technology research in the medical sciences (Table 1.11).

Table 1.11 Country productivity for robotics research in the medical sciences

h-index	Country name	h-core citation sum	All citations	All articles
146	United States	39 042	129 513	4 661
52	United Kingdom	4 677	10 261	601
50	Germany	6 468	11 760	648
48	Italy	5 931	12 277	680
42	France	3 829	8 796	660
39	Canada	4 448	8 788	516
38	South Korea	4 628	7 789	450
35	Netherlands	4 469	5 621	150
34	People's Republic of China	4 405	6 998	548
33	Japan	3 905	7 012	576
32	Switzerland	3 379	4 725	164
29	Australia	2 461	3 537	179
27	Spain	2 334	3 203	185
23	Singapore	1 633	2 130	122
23	Belgium	1 732	1 986	66
23	Israel	2 016	2 329	91
22	Taiwan	1 032	1 999	240
21	Greece	953	1 230	90
20	Austria	963	1 200	71
18	Sweden	785	1 053	58

Funding dynamics

Examining the funding status of robotic technologies in the health sciences, as a field of research and technology as it manifests in our study, we observed that the first study to receive funding was published in 1999. The publications had an average of one source of funding until 2005, a number which increased and peaked in 2008 with an average of 2.07. The studies that received the most funding support received support from 14 different fund providers.

Table 1.12 Source dynamics for robotics in the medical sciences

h-index	Journal name	h-core citation sum	All citations	All articles
30	*IEEE Transactions on Robotics*	2 144	2 468	58
29	*International Journal of Robotics Research*	2 995	3 240	45
25	*Surgical Endoscopy and Other Interventional Techniques*	1 709	2 320	83
25	*Journal of Urology*	1 671	1 934	46
25	*International Journal of Medical Robotics and Computer-Assisted Surgery*	1 300	1 930	102
23	*IEEE/ASME Transactions on Mechatronics*	1 447	1 865	65
21	*Medical Physics*	1 355	1 837	106
19	*BJU International*	940	1 077	29
19	*European Urology*	1 142	1 168	22
19	*American Journal of Surgery*	1 291	1 368	28
18	*Journal of Endourology*	938	1 167	45
18	*IEEE Transactions on Biomedical Engineering*	873	1 039	38
17	*Archives of Physical Medicine and Rehabilitation*	1 516	1 615	34
16	*Fertility and Sterility*	719	775	27
15	*Urology*	841	940	35
15	*Journal of Minimally Invasive Gynecology*	462	622	49
15	*Proceedings of the IEEE*	762	771	16
13	*Gynecologic Oncology*	574	620	22
12	*IEEE Robotics and Automation Magazine*	641	664	14
12	*Industrial Robot*	872	924	34

Source dynamics (journals, proceedings, and so on)

The journals in which studies on robotics technologies and MS were frequently published are given in Table 1.12. When the source dynamics were analyzed, it could be observed that the top two journals in which the articles were most commonly published were journals of robotic technologies, and the journals that followed were specialized in the field of surgery.

Collaboration

In our study, co-occurrence analysis was used to determine cooperation patterns. As in the productivity analysis, collaboration analyses were conducted at the author, institution, and country levels. As we mentioned previously, the analysis of collaborations is among the methods used to identify individuals, institutions, and organizations engaged in joint studies in a scientific discipline or subject area. With this analysis, which is called co-word analysis, it is possible to access data in support of technology and engineering management. In this part of the study, we used a keyword plus co-word analysis to determine the themes of robotics research in the medical sciences. In this way, we identified what is popular in the field. As a result of the co-word analysis, it was found that robotics research in the medical sciences focuses on diagnostic purposes, while software, hardware, and method research was aimed at assisting surgery (Figure 1.7).

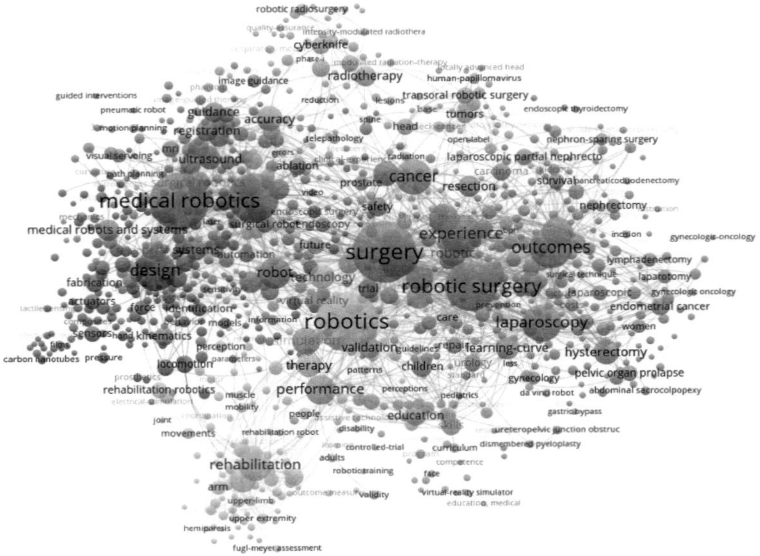

Figure 1.7 *Co-word analysis of keyword source dynamics for robotics in the medical sciences*

Using the figure showing the network values, it is possible to summarize the prominent concepts as follows: surgery, system, experience, and surgical robots (Table 1.13).

Institutional collaborations

When the data are analyzed in terms of institution productivity, it is possible to derive a table that is similar to the country productivity table. Five US universities are at the top of the list, while the sixth is the Mayo Clinic, a non-profit organization. When the institutions carrying out robotics research in MS and the business intelligence they have obtained are examined, it can be seen that robotics is considered, as an emerging technology area, as a more capped

Table 1.13 *Co-word analysis of keywords for robotics in the medical sciences*

Label	Links	Total link strength	Occurrences	Average publication year	Average citations	Normalized average citations
Surgery	701	2 960	570	2012	194.474	1.112
Robotics	666	2 542	555	2011	191.766	0.9817
System	562	1 961	361	2012	226.898	12.661
Robotic surgery	532	1 888	369	2013	144.688	10.495
Experience	463	1 460	234	2012	226.752	11.795
Medical robotics	445	1 866	483	2011	195.445	0.9662
Minimally invasive surgery	440	1 159	228	2012	168.904	0.9527
Outcomes	407	1 551	251	2014	158.526	11.521
Cancer	392	1 050	188	2013	196.277	11.675
Design	372	1 026	228	2014	255.439	23.089
Laparoscopy	369	1 326	224	2013	190.446	10.673
Performance	322	791	127	2013	236.535	13.479
Robot	321	633	144	2012	129.444	0.8338
Complications	293	837	134	2014	202.463	15.395
Management	293	625	127	2013	17.126	10.235
Impact	280	567	85	2013	219.529	12.677
Robotic	278	555	105	2013	154.857	10.059
Laparoscopic surgery	270	594	110	2011	220.455	11.033
Therapy	269	525	87	2012	347.241	14.652
Feasibility	267	600	94	2012	201.809	10.796

ecosystem; in other words, institutions prefer to develop collaborations within their own departments (Figure 1.8). In terms of collaboration, as also seen in Chapter 2, the tendency of research institutions to collaborate with the subunits within their own institutions can be attributed to the high commercial potential of the new technologies. In order to better understand this structure of collaboration, which we can define as introverted, it is useful to discuss in depth such methods as technological knowledge redundancy.

It can be observed that universities with Chinese addresses form a separate cluster, and institutions with Japanese addresses also form a cluster with each other. Similarly, it can be seen that the research institutions of the EU member countries form a cluster with each other. It is possible to list the individual prominent institutions. When we look closely at the values for collaboration, Johns Hopkins University, Harvard University, Stanford University, Harvard

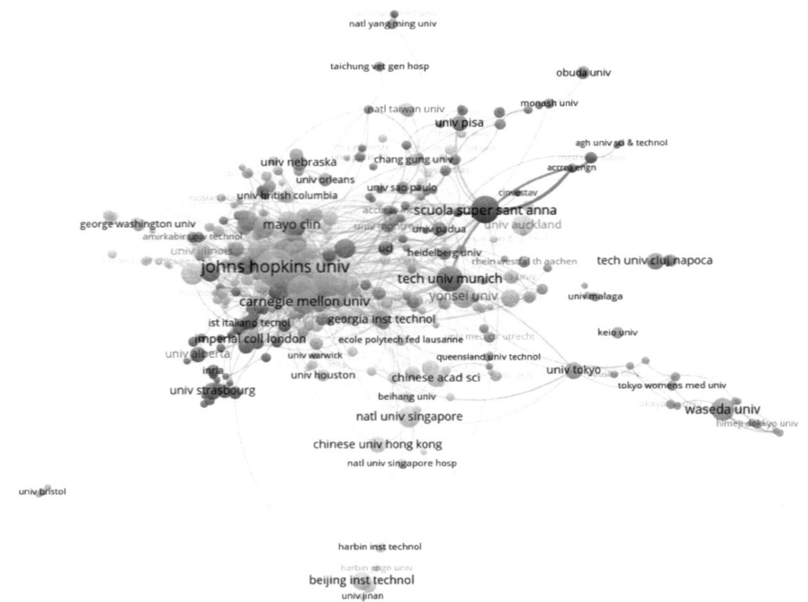

Figure 1.8 *Institutional collaborations in robotics in the medical sciences*

Medical School, and the Mayo Clinic are clearly the institutions most prone to collaboration. This situation can be interpreted in terms of collaboration having a positive effect on productivity as it is parallel to the values for the corporate productivity measures (Table 1.14).

It can be seen that research is concentrated on surgical technologies at the development phase. It is noteworthy that institutions chose to collaborate within their own departments. As robotics research is an emerging technology, it should be noted that the observation of such a collaborative partnership is natural. Such a closed pattern in terms of collaborations prevents the attainment of the desired level of development in robotics research. In this context, it can be said that it has a destructive effect on the known business models. Among the main objectives of TM are the identification of the methods used and the datasets applied in predecessor and successor technologies. The literature-based discovery method is one of the leading methods used for this purpose. When bibliometric methods are applied, it is possible to make inferences about vital points such as the determination of the dominant actors and of the concentration of scientific domains of a related technology. Predictions

Table 1.14 Institutional collaboration network parameters for robotics in the medical sciences

Institution name	Links	Total link strength	Documents	Citations	Normalized citations	Average publication year	Average citations	Normalized average citations
Johns Hopkins University	83	150	123	4377	2 036 042	2012	355 854	16 553
Harvard University	66	121	83	3668	2 269 214	2012	441 928	2 734
Stanford University	63	105	79	5544	3 492 484	2011	701 772	44 209
Harvard Medical School	63	91	31	388	744 737	2015	125 161	24 024
Mayo Clinic	48	73	44	1214	1 003 955	2013	275 909	22 817
Duke University	52	72	36	760	462 373	2013	211 111	12 844
University of Pittsburgh	44	72	64	2541	1 202 238	2011	397 031	18 785
University of Michigan	50	71	36	638	645 177	2014	177 222	17 922
Massachusetts Institute of Technology	34	68	58	2861	1 450 344	2011	493 276	25 006
Brigham and Women's Hospital	37	66	34	1036	797 145	2014	304 706	23 445
Vanderbilt University	42	64	62	1471	1 110 373	2013	237 258	17 909
University of California Irvine	48	62	27	1314	877 085	2011	486 667	32 485
Cleveland Clinic	43	61	67	1293	1 191 742	2015	192 985	17 787

Institution name	Links	Total link strength	Documents	Citations	Normalized citations	Average publication year	Average citations	Normalized average citations
Carnegie Mellon University	30	59	54	2232	1 516 461	2012	413 333	28 083
University of California, San Francisco	42	58	23	791	423 188	2011	343 913	18 399
University of Miami	36	54	17	674	490 297	2012	396 471	28 841
Imperial College London	41	53	37	378	545 452	2015	102 162	14 742
University of Chicago	36	52	21	647	511 213	2012	308 095	24 343
Northwestern University	32	51	40	1084	645 593	2012	44 223	1 614
Case Western Reserve University	32	49	34	1112	597 322	2012	327 059	17 568

can be made about when the studies related to the technology will reach a saturation point using the growth curves for the related applications. The determination of the sub-technology domains to be invested in for the technologies that have passed from the basic research level to the product research level is among the issues examined in this context.

Discussion

The use of robots in medical fields emerged as a phenomenon in the 1980s (Gourin & Terris, 2007). While the first robots offered surgical assistance through robotic arm technologies, as a result of the recent advances, AI-supported image processing and data analysis techniques have also contributed to the development of medical robots. Today, robotic technologies are used not only in surgeries but also in clinical settings for activities such as supporting healthcare professionals and patient care.

When the dataset was examined, it could be observed that the number of studies on robotic technologies had increased since the 1980s. We deduced that robotics aimed at facilitating medical applications are intensively integrated into the surgical field in the medical sciences. Robotics technology aims to reduce the errors caused by human intervention by developing robotic operations involving robotic arms or computers under the control of surgeons. It was observed that integration with other devices used in medical interventions is an important issue, especially in complex areas such as microsurgery, where the benefits and advantages of long-lasting, open, and compelling operations are known to shorten the duration and reduce the costs of operations. Robotics has contributed to the development of methods that are important in cancer treatment and diagnosis, biopsies, and operations.

A SCIENTIFIC REVIEW OF THE LITERATURE ON TRANSGENIC FISH

The rapid development in the field of biotechnology has enabled the chromosomes and genes of living organisms to be manipulated (Dunham, 2011; Zbikowska, 2003). Today, transgenic fish have been obtained by using transgene technology, biology, medicine, veterinary medicine, animal husbandry, agriculture, pharmaceuticals, and so on. They are used in the large-scale production of different molecules for scientific research models, therapy, and diagnosis purposes (Chen & Powers, 1990; Liu et al., 1990). In terms of aquaculture, transgenic studies are carried out with the aim of obtaining fish that grow faster than normal and are tolerant of diseases, low water temperatures, and low levels of oxygen (Boleti et al., 2020; Chen et al., 2020; de Araújo Dunham, 2011; Hua & Ekker, 2020). The genetic modification of fish aims to improve the biological structure of particular fish species, especially to increase the resistance of the fish (Akhan & Canyurt, 2008; Ekici et al., 2006; Özden et al., 2003). A living being whose genome has been transferred from another organism using biotechnological methods is called 'transgenic'. Transgenic research has been carried out on a variety of living beings, from plants to animals. In transgenic research on plants, the origin of the gene transferred can be a plant belonging to another plant species that cannot be hybridized with the plant to which the gene is transferred, or in research on animals the focus may be in the form of animal color change.

We examined transgenic fish research activities using a bibliometric analysis of the international literature, as it is one of the most commonly used fields of technological research today. In this context, in order to compile the bibliographic data for transgenic fish research as reflected in the international literature, a topic search was applied in all relevant databases of WoS. The obtained data were made uniform and recorded in a relational database in order to prepare them for analysis. Descriptive statistics and basic bibliometric indicators related to transgenic fish research are given in Table 1.15.

Table 1.15 Descriptive statistics for the data

Timespan	1986–2020
Documents	782
Average years from publication	12.4
Average citations per document	32.01
Average citations per year per document	2.419
References	23 163
Authors	2 462
Author appearances	3 867
Authors of single-authored documents	47
Authors of multi-authored documents	2 415
Author collaborations	
Single-author documents	53
Documents per author	0.318
Authors per document	3.15
Co-authors per document	4.95
Collaboration index	3.31

Although the number of studies with a single author was quite low, it was observed that the cooperation index for transgenic fish studies was quite high. The ratio of the number of citations per document was also remarkable (32.01).

When looking at the annual growth rate of the research literature on transgenic fish applications, it is worth mentioning that it is a developing technology with a rate of approximately 10 percent (9.64 percent). The results for the contributions of countries to this technology in terms of the numbers of publications, the subject areas, and the institutions are presented in Figure 1.9. Accordingly, while the US and Japan are in the lead, countries such as China and Canada have made remarkable contributions in terms of the numbers of publications.

The SNA method was used to determine the countries at the key points in transgenic fish research. In this respect, it was found that the US, Canada, and China are at the center of such research. It should be stated that, in terms of centrality values, the UK, Japan, Germany, and France also hold central positions for transgenic fish research. Countries such as Turkey and Peru still act in cooperation with their own national institutions and organizations and can be observed to draw patterns independently of the center of the network (Figure 1.11).

When the contributions of the countries are examined in terms of the effects they have created in the literature, a table showing similarities with the values seen in the network map is formed. It should be noted that while the US leads the list in terms of both the number of publications and the number of citations, Japan follows the US in terms of the number of publications and the number of citations it receives. It is worth noting that France, which is located close to the center of the transgenic fish research network in terms of network values, falls behind many countries in terms of both the number of publications and the number of citations. In this respect, it is obvious that in-depth analysis of the contributions of research institutions and organizations in parallel with the contributions of countries would be a very efficient tool, especially in determining experts in TM. For this purpose, we determined the leading research institutions in transgenic fish research using an analysis of the address information for the data we had (Table 1.16).

In order to examine the contributions of countries in more depth, affiliation analysis was performed to identify leading institutions and organizations, and pioneering research institutions in the field of transgenic fish research were identified. Thanks to the list that was used, the leading institutions in the field could be identified and the effects they have on the field revealed. In other words, related institutions could be compared in terms of both the numbers of publications they were part of and the numbers of citations they received in the field (Table 1.17).

When the important research institutions in the field of transgenic fish are examined, it can be seen that Fisheries and Oceans Canada is at the top of the list in terms of h-index values, which represents the intersection point of publication and citation performance, regardless of the performance of the countries, while the individual performances of the institutions with Japanese

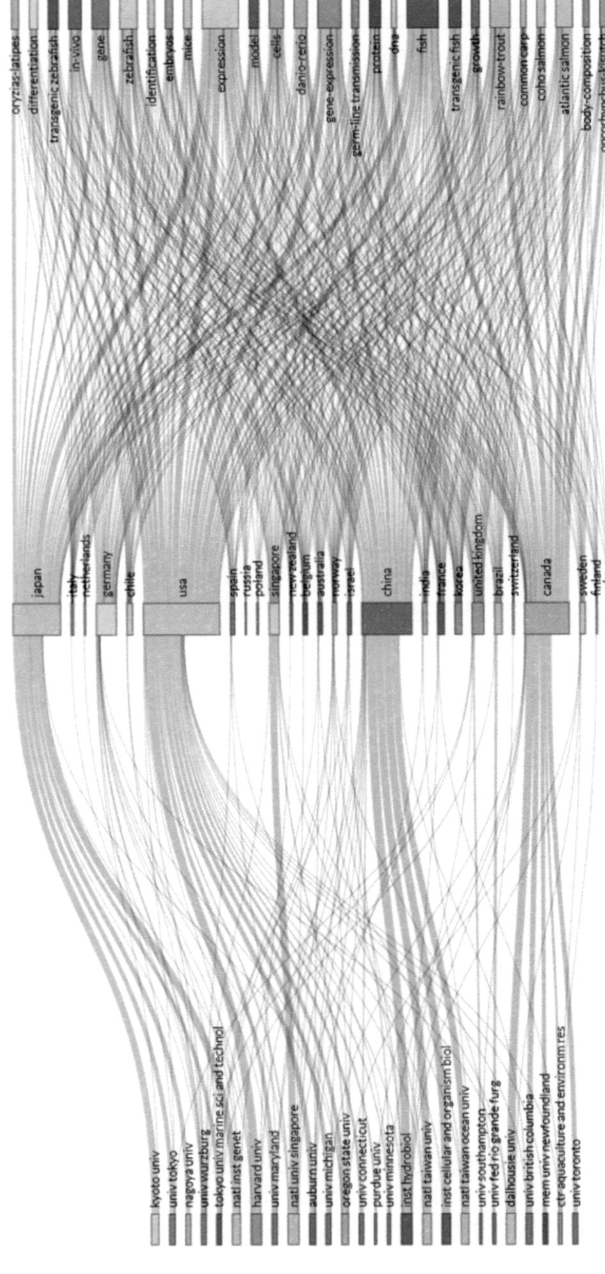

Figure 1.9 *Countries, institutions, and keywords*

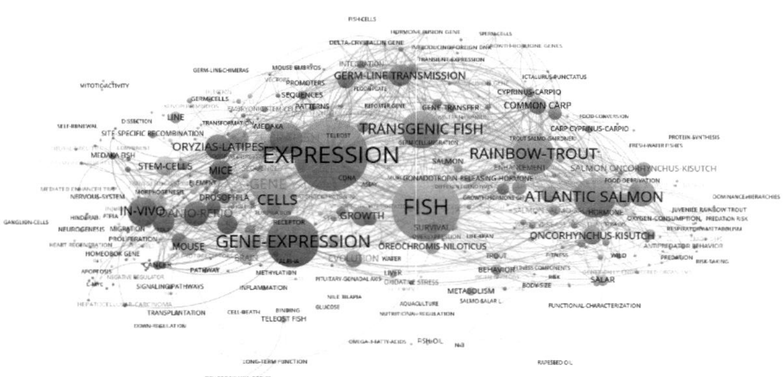

Note: The network of keywords and words frequently used in the field of transgenic fish research is visualized in Figure 1.10. When looked at closely, it can be seen that transgenic fish research is clustered around six main clusters: gene identification studies; medical studies; fish species and behavior; cell studies; genes and biology; and procedures.

Figure 1.10 Keyword-Plus co-concurrency analysis

addresses are remarkable (National Institute of Genetics, Tokyo University of Marine Science and Technology, University of Tokyo, and Kyoto University). Examining the collaboration between research institutions in detail reveals that the partnership patterns in transgenic fish research can be used to determine the main stakeholders of this technology and to determine the geographic regions of possible technology foci. As seen in the co-occurrence analysis conducted for this purpose, researchers exhibited a pattern of working with other researchers in their institutions, while patterns of collaboration could be seen for the National University of Singapore and Oregon State University, Indiana University School of Medicine and Indiana University School of Medicine, Indiana University School of Medicine and the University of Chicago, Fisheries and Oceans Canada and the University of British Columbia, National Taiwan Ocean University and Academia Sinica, and Children's Hospital and Harvard University (Figure 1.12).

Discussion

After the Second World War, the rapid increase in the global population brought about the need for brand new food resources. The first response to this need was the 'Green Revolution' in the late 1960s. However, over the years, the serious side effects of this highly hopeful initiative came to attention and

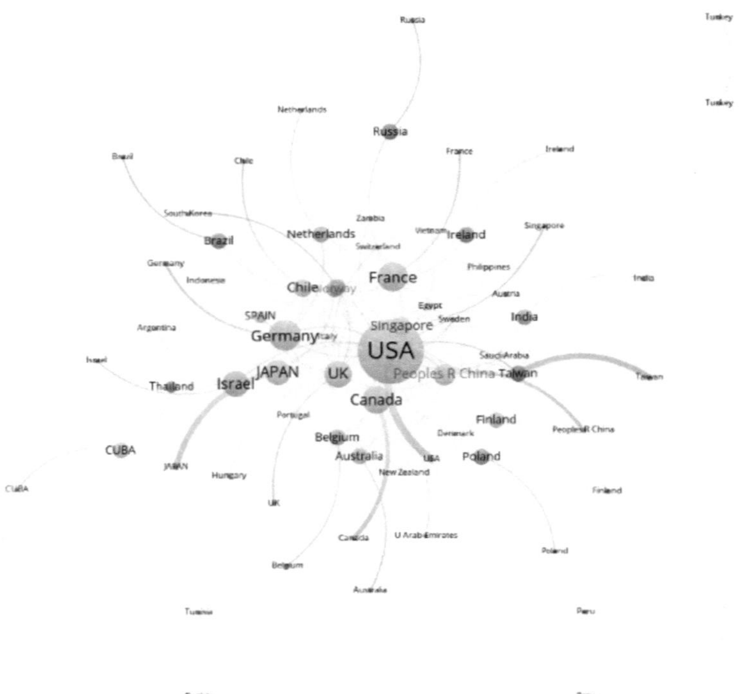

Figure 1.11 Country network for transgenic fish research

many of the chemicals that were used were banned in the 1970s (Brundtland, 1987). The continued search for new food resources catalyzed the use of the gene transfer technology developed at that time to increase agricultural productivity.

R&D is carried out on transgenic species for reasons arising from various needs. Regardless of the reasons, it is necessary to examine research on any technology with an approach that explores all dimensions. Emerging technology has gained importance in biotechnology research and found a place for itself. Although positive developments in biotechnology studies bring results that positively affect the quality of human life, it is necessary to consider such studies from social, ethical, economic, and environmental perspectives. In other words, it is beneficial to consider such technology R&D studies from multiple perspectives and to examine the possible results just before technology investments are made. In this context, in addition to research on transgenic blockages and transgenic species, it is necessary to address the issue from social, cultural, and economic perspectives in the context of technology management.

Table 1.16 Country productivity for transgenic fish research

h-index	Country name	h-core citation sum	All citations	All articles
74	United States	11 432	19 721	473
59	Japan	9 392	12 933	279
49	Canada	4 682	7 628	221
31	Germany	2 798	3 334	83
28	Taiwan	1 458	2 405	140
28	United Kingdom	2 045	2 352	59
25	People's Republic of China	1 124	2 717	177
20	Singapore	1 246	1 563	49
19	France	1 499	1 598	40
16	Sweden	965	1 027	27
14	Norway	800	828	21
14	South Korea	722	816	31
14	Brazil	322	488	39
14	Australia	504	580	23
14	Israel	608	657	20
11	Spain	299	358	23
11	Chile	518	559	23
9	Cuba	448	448	9
8	Switzerland	332	332	8
8	India	105	141	18

Table 1.17 Institution productivity for transgenic fish research

h-index	Institution name	h-core citation sum	All citations	All articles
24	Fisheries and Oceans Canada	1 583	1 928	67
21	Harvard University	1 873	1 971	27
20	Chinese Academy of Sciences	639	1 165	68
19	National University of Singapore	957	1 108	38
17	National Institute of Genetics	2 191	2 227	20
16	University of Southampton	789	796	17
16	Dalhousie University	466	521	24
15	University of Minnesota	703	716	17
14	University of Michigan	1 217	1 243	19
14	Oregon State University	960	996	19
13	University of Toronto	1 219	1 243	15
13	Academia Sinica	390	469	29
13	University of Maryland	696	716	17
13	Memorial University of Newfoundland	655	688	17
13	Tokyo University of Marine Science and Technology	401	446	21
12	University of Tokyo	1 087	1 149	23
11	Johns Hopkins University	1 368	1 372	12
11	National Taiwan University	347	400	22
11	Kyoto University	435	478	21
11	University of British Columbia	481	518	21

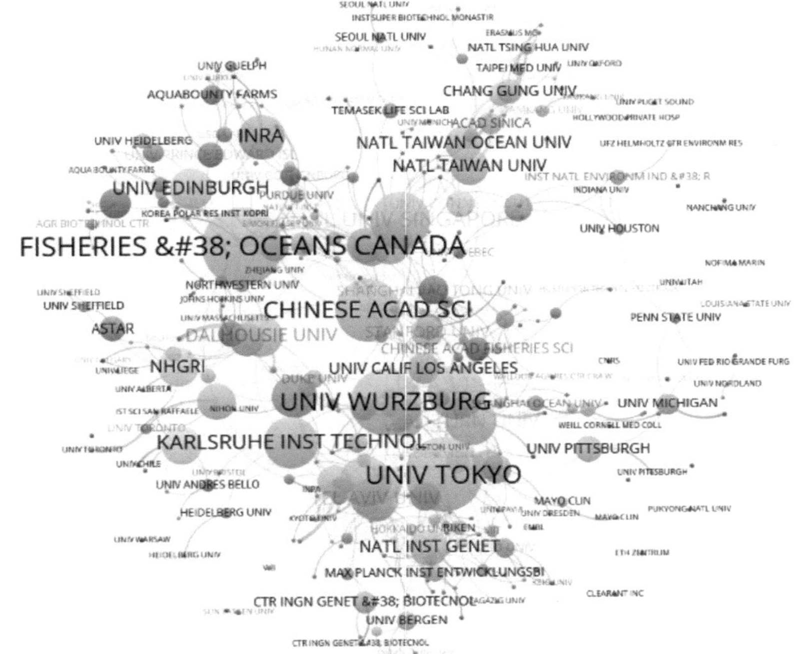

Figure 1.12 Institution collaboration patterns in transgenic fish research

NOTES

1. TS=('coronavirus') or TS=('severe acute respiratory syndrome') or TS=('coronaviridae') or TS=('SARS-CoV') or TS=('MERS-CoV') or TS=('COVID-19') or TS=('Wuhan coronavirus') or TS=('2019-nCoV') or TS=('Middle East respiratory') or TS=('SARS') or (TS='MERS').

2. Patent-based analyses

Technological improvements can be monitored using patents. For example, the relations between technologies can be determined by using patent citations (Huang et al., 2003; Moed et al., 2004; Zhou et al., 2014). Nowadays there are huge numbers of patents for any technological domain, so it is difficult to manually read and analyze all patent data for patent intelligence. The analysis process needs an automated system to analyze and interpret the patents with given criteria (Tseng et al., 2007). Patent automation systems play a key role in the market. In the past, companies would hire patent experts to follow innovations and trends. Although various automation systems have been developed for use in patent analysis today, it is not possible to say that there is a one-size-fits-all solution that meets every institution's features and expectations. In this respect, it is vital to design patent analysis in accordance with the expectations of institutions. A traditional patent analyzing system deploys the following steps:

- Task identification: identify the scope of the search;
- Searching: search, filter, and download patents;
- Segmentation: pre-processing step before analysis using text-mining techniques to convert unstructured data to structured data;
- Abstraction: analyze the structured data (claim, abstract, description) and summarize them;
- Clustering: group the patents based on some of the extracted features;
- Visualization: create meaningful images to monitor the clustering results;
- Interpretation: identify and interpret technology trends and relations.

Patents are a core source for analyzing the technology-related activities of a specific technology domain. Many studies can be undertaken using patent data. Patents not only contain related technology information but also involve information referring to other technology fields.

First, it is necessary to use more than one method to produce solutions for all the levels required in patent analysis. In this context, TM stands out. However, before applying TM principles to patent data, it is necessary to understand the patent data correctly. After that, TM can be used to extract and analyze that information effectively. Before processing a patent, its internal structure should be fully understood. The internal structure provides a lot of information and various clues about what kind of information can be extracted. Along with text-mining technologies, TM plays an important role in analyzing this information (Madani, 2015). The TM method was first introduced in 2004 (Porter & Cunningham, 2004) and is still applied by various researchers in different research domains, such as medicine, biology, nanotechnology, computer science, and so on. It is a subdomain of R&D and also provides opportunities to capture technological innovations and edge out competitors. Since it was introduced, TM has been used in many fields. However, there are also sectors where TM has never been used because it has traditionally been considered an expensive investment.

MINING TECHNOLOGICAL INVENTIONS AND THE R&D LANDSCAPE OF BIOPRINTING TECHNOLOGIES IN MEDICAL APPLICATIONS

In this study, the patents in the field of bioprinting technologies were analyzed. The kind of support that the obtained data can provide to decision makers in the context of technology and engineering management has been shown. The methods and tools used in patent analysis differ. In this study, a combined application of text mining, data mining, bibliometrics, and SNA methods, which we can define as TM for the methods used in patent analysis, is used to present examples in the patent analysis.

Patent Analysis

Patent analysis is one of the most frequently used methods for TM. Thanks to this method, it is possible to identify which technology domain is the research focus. On the other hand, patent analysis is frequently used to determine the direction (rising or falling) of a technology trend and to conduct competitor analysis (Feng et al., 2021; Wilson, 1987). While companies can guide their strategic planning activities thanks to the data obtained from patents, they also make it possible to identify pioneers in a technology transfer domain. It is further possible to identify the frequently cited patents, thereby forming the basis for the production of new ideas or for an analysis that covers the most important work in the relevant technology (Cho et al., 2021). Patent analysis can also be used to decide in which countries a specific technology should be protected (Schröder & Widera, 2021). With patent analysis, potential customers who are interested in a specific technology can also be identified. Patent analysis is very important for managers in the decision-making process as it can provide strategic data regardless of the sector studied (Jun et al., 2021). The method is also frequently addressed in academic studies. Determinations of specific technologies in these studies (Mu-Hsuan & Hsiao-Wen, 2013) take the form of methodological improvement efforts (Daim et al., 2006; Zhou et al., 2019) or case reports (Rajeswari, 1996). Therefore, it is possible to say that such applications have been addressed. Biglu (2009) reported that the method was used frequently between 1965 and 2005. Patent analysis combined with bibliometrics can also be used to determine the particular geographic areas where technologies might potentially be developed. An example of this kind of study is the work carried out by Pouris in 1991 in order to determine the technologies owned by South Africa (Pouris, 1991). An examination of the literature shows that there are studies that aim to determine the structure of the biotechnology field (McCain, 1995), while other studies aim to reveal the structure of interdisciplinary technologies (Hinze & Grupp, 1996). Some studies have used patent analysis to better understand the general characteristics of the original scientific domains (Meyer, 2000, 2001). There are studies that focus on the use of indicators produced from competition management and owned data in developing regional technology policy (Coronado et al., 2004; Ramani et al., 2001; Verbeek et al., 2001). Patent analysis has found a new use area and has become a basic tool used in determining the technology policies of both companies and countries (Cavalheiro et al., 2021). Over time, the strategic data provided by patent analysis has led to the privatization of such studies as a technology policy monitoring tool (Pao-Long & Hoang-Jyh, 2010). Following the developments in computer and information technologies and increases in computing capacity, it is possible to combine text mining and patent analysis and examine pattern determination processes through methods such as subject determination and cluster analysis using the full texts of patents (Vicente-Gomila et al., 2017; Wittfoth, 2019; Yeh et al., 2018; Zhou et al., 2019).

Bioprinting and Previous Studies

Patent analysis is an indispensable tool of technology and engineering management. In this section, in which we examine the ways of handling three-dimensional (3D) printer technologies in the medical sciences, we aimed to present a case of how patent analysis can be used in technology management. Examining the literature, it can be seen that the number of studies on the subject discussed here is limited (Choudhury et al., 2018; Gopinathan & Noh, 2018). While it can be observed that the parts produced for people entail ethical concerns and that the legal dimensions are intensified, authors have provided warnings that these issues will be affected by the legal infrastructures of countries (US) or regions (European Union, etc.) (Ebrahim, 2017; Minssen & Mimler, 2016). In their approach, Rodríguez-Salvador et al. conducted an analysis of 3D printer technologies using a hybrid method (Rodríguez-Salvador et al., 2017). The main results of the study were that bioprinting technology will have a groundbreaking effect in tissue engineering as a disruptive technology. This result is similar to the one obtained by Choudhury et al. (2018; Gopinathan & Noh, 2018). Although the analysis of studies is an important example in terms of technology management, it can be improved using SNA. Therefore, bibliometrics and SNA methods are here handled together to determine the technology domains where 3D printer technologies are present as emerging technologies.

Method and Data Collection Technique

The Derwent Innovation Index (DII) was used for the dataset in our study. Patents published in the field of 3D printing between 2008 and 2020 were obtained. Since the scope of our study only included patents in the medical sciences, the data obtained were refined according to this information. After this refinement, the total number of patents analyzed was 2098. To determine the distribution of patent ownership for 3D printer technologies in the medical sciences, we concentrated on determining which sub-technology domains the patent registrations were associated with. By considering the bibliometric analysis and the SNA, the sub-technology domains related to 3D printer technologies were visualized. Our goal was to create a user guide for technology managers dealing with this technology domain using the data obtained.

Basic Bibliometric Indicators

Considering the publication years, it was observed that there has been a significant increase in patent registration processes for 3D printing technologies since 2006. It can be stated as an important finding that registrations increased by more than two and a half times in 2016 compared to the previous year. Patent registration processes accelerated radically in 2016 (Figure 2.1).

The tag cloud produced for the words used in the titles of the patented ideas is presented in Figure 2.2. In order to determine the intensity of technology ownership, an analysis was carried out on patent application ownership notifications. Accordingly, it was concluded that ownership is realized both at the level of real persons and that of legal persons, while in patent ownership and, therefore, in R&D studies, partnerships are formed using a limited number of relationships within the same institutions. Considering that this is a period in which we see the early processes of research, it can be said that this is an expected development. By conducting a classification study to determine the technology domains of the registration processes, it is possible to determine the technology domains that reach the saturation level, if any.

Details of the top 20 patents for bioprinting with the highest levels of technology ownership are presented in Table 2.1.

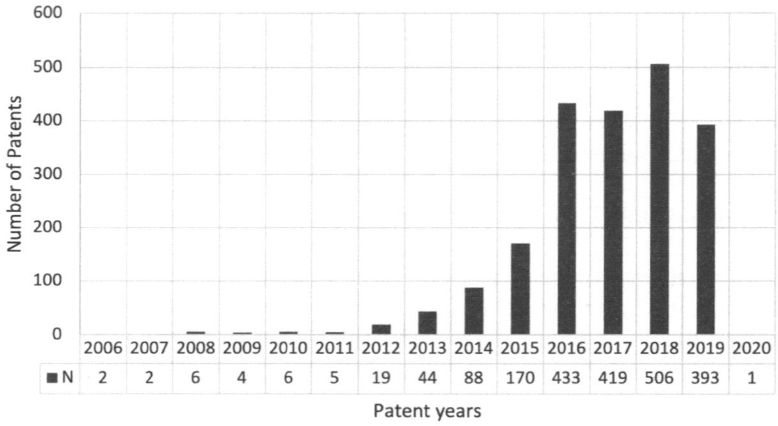

Figure 2.1 Bioprinting patent documents by year

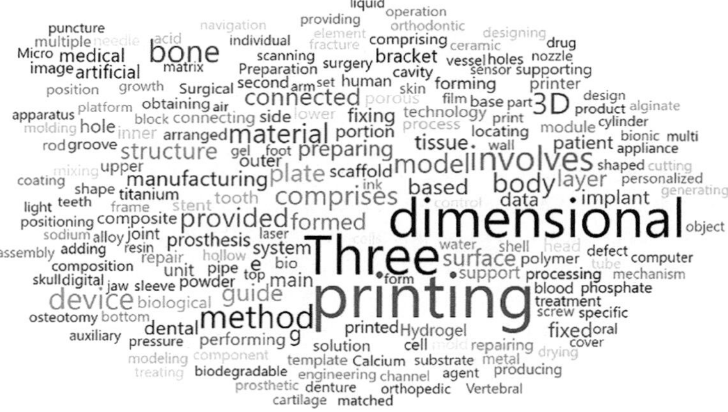

Figure 2.2 Tag cloud (title terms)

It is possible to see that 3D printer technologies are more frequently associated with the dental health sector and the optical domain. In other words, the sectors where registration is most intense are dental materials and optics. Considering that R&D studies can turn into commercially important outputs, this pattern is quite understandable. When we examine the table closely, it can be observed that the property rights of the ideas registered in the field of bioprinting belong to institutions from Asia. The technology domains in which the patents are

Table 2.1 Patent assignees for bioprinting

No.	Assignee	Patents owned
1	Shenzhen Jiahong Oral Medical Company Ltd (SHEN-Non-standard)	8
2	Beijing Unicom Science and Technology Company Ltd (BEIJ-Non-standard)	8
3	Hunan Huaxiang Incremental Manufacturing Company Ltd (HUNA-Non-standard)	7
4	Shanghai Hanhua Dental Materials Company Ltd (SHAN-Non-standard)	6
5	Shaanxi Doworld Technology Company Ltd (SHAA-Non-standard)	6
6	Hunan Liuxin Intelligent Technology Company (HUNA-Non-standard)	6
7	Peking University Third Hospital (UYPK-C)	6
8	Liu, R (LIUR-Individual)	5
9	Yunnan Marvel Technology Company Ltd (YUNN-Non-standard)	5
10	Suzhou Hongjian Automation Technology Company (SUZH-Non-standard)	5
11	West China Hospital of Sichuan University (USCU-C)	5
12	Children's Hospital of Nanjing Medical University (UYNM-C)	5
13	Shanghai Jiao Tong University School of Medicine – Ninth People's Hospital (USJT-C)	4
14	Shaanxi Hengtong Intelligent Machine Company (SHAA-Non-standard)	3
15	Henan Diaoxin Technology Company Ltd (HENA-Non-standard)	3
16	Yang, C (YANG-Individual)	3
17	Hkable 3D Biological Printing Technology (HKAB-Non-standard)	3
18	Hangzhou Regenovo Biological Technology (HANG-Non-standard)	3
19	Xiamen No. 5 Hospital (XIAM-Non-standard)	3
20	Shanghai Children's Medical Center (UNIV-Non-standard)	3
21	Xinjiang University (UXIJ-C)	3

registered can provide us with useful information for the determination of which technology domains are the research focus. In this regard, manual codes, which are among the bibliographic data of patent publications, were evaluated. Social network maps, where the same classification codes were created by considering the frequency of patents' use together, were used to see the sub-technologies bioprinting patents are classified under (Figure 2.3).

As can be seen in the results of the co-word analysis, the A96 ('Medical, dental, veterinary, cosmetic') cluster represented the most intensely registered technology domain of its class. On the other hand, this cluster was most frequently related with the X25 ('Industrial electric equipment') cluster. In other words, another technology domain under which bioprinting ideas, which are heavily patented in the medical, dental, veterinary, and cosmetic fields (A96), are frequently classified is industrial electric equipment, in the X25 class. When evaluated from this point of view, it is possible to say that the need to patent the applications produced with bioprinting technologies as industrial applications has become a priority (Table 2.2).

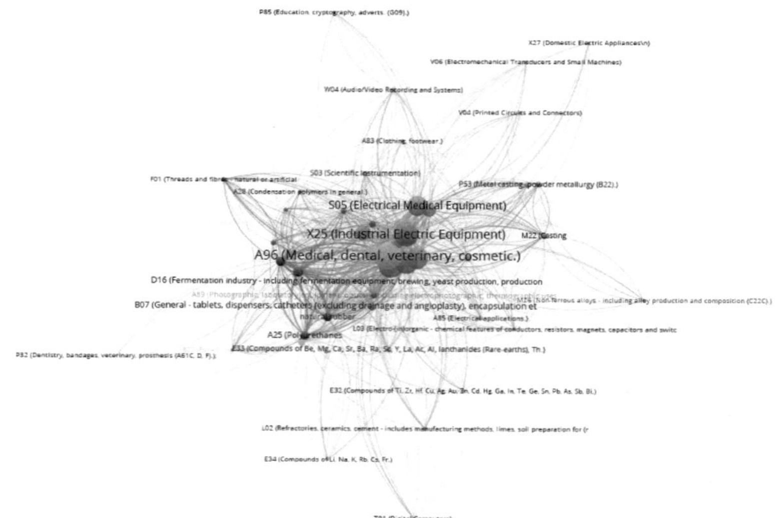

Figure 2.3 Manual code classifications (bioprinting)

A circular dendrogram visualization was used to better understand the relationships between patent classifications and to follow the nodes overlapping on the map obtained with the SNA. In this way, we can see the sub-technology domains and sub-technology classes more easily (Figure 2.4). When Figure 2.4 is examined, it is possible to see that the X25 technology class is classified together with many other technology subclasses. It is clear that an in-depth study of this image can provide data and shed light on the identification of other subdomains of bioprinting patents that are open to improvement.

The registration of patents in various countries is a testament to the extent of the intellectual property rights of the idea. In this regard, a list of the most comprehensive patents according to the number of countries where they are protected is given in Table 2.3 below. The protection of the property rights of the registered commercial ideas is carried out with the aim of obtaining the highest possible commercial income from the R&D activities. Since this also registers the originality of the idea, this indicator means that a patent can be perceived at a high level in terms of its technological originality or novelty. In this regard, it can be said that the number of regions where patents are registered is an important indicator of the originality and novelty of a technology.

Table 2.2 *Patent classification network parameters (bioprinting)*

Label	Cluster	Links	Total link strength*
A96 (Medical, dental, veterinary, cosmetic)	4	55	3 416
X25 (Industrial electric equipment)	2	54	2 764
D22 (Sterilizing, bandages, dressing and skin protection agents …	3	54	2 623
P34 (Sterilizing, syringes, electrotherapy (A61L, M, N))	3	53	2 374
P32 (Dentistry, bandages, veterinary, prosthesis (A61C, D, F))	2	54	2 271
S05 (Electrical medical equipment)	2	48	1 938
T01 (Digital computers)	2	53	1 744
A32 (Polymer fabrication, such as moulding, extrusion, forming, laminating, spinning)	2	47	1 031
B04 (Natural products and polymers. Including testing of body fluids…	4	53	949
P31 (Diagnosis, surgery (A61B))	2	44	818
B07 (General – tablets, dispensers, catheters (excluding drainage and angioplasty), encapsulation…	4	42	600
S06 (Electrophotography and photography)	2	47	508
A14 (Polymers of other substituted monoolefins including PVC, PTFE)	1	39	503
Polyamides; polyesters (including polycarbonates, polyesteramides); alkyds; other unsaturated polymers	1	37	453
D16 (Fermentation industry – including fermentation equipment, brewing, yeast production, production…	4	39	363
A11 (Polysaccharides)	1	37	362
A25 (Polyurethanes)	1	39	353
D21 (Preparations for dental or toilet purposes, including filling alloys, compositions for dentures or dental impressions, anti-caries chewing gum, plaque-disclosing compositions, toothpastes, cosmetics, shampoos, topical anti-sunburn compositions, and toilet soaps (A61K))	3	41	265
A97 (Miscellaneous goods not specified elsewhere, including papermaking, gramophone records, detergents, food, and oil well applications)	1	39	242
B05 (Other organics – aromatics, aliphatic, organo-metallics, compounds whose substituents vary such that they would be classified in several…	4	30	218

Note: *The number of links of an item with other items and the total strength of the links of an item with other items.

Discussion

The bioprinting patent analysis clearly showed us that the research dynamics in the medical sciences are focused on the dental health and optics sectors.

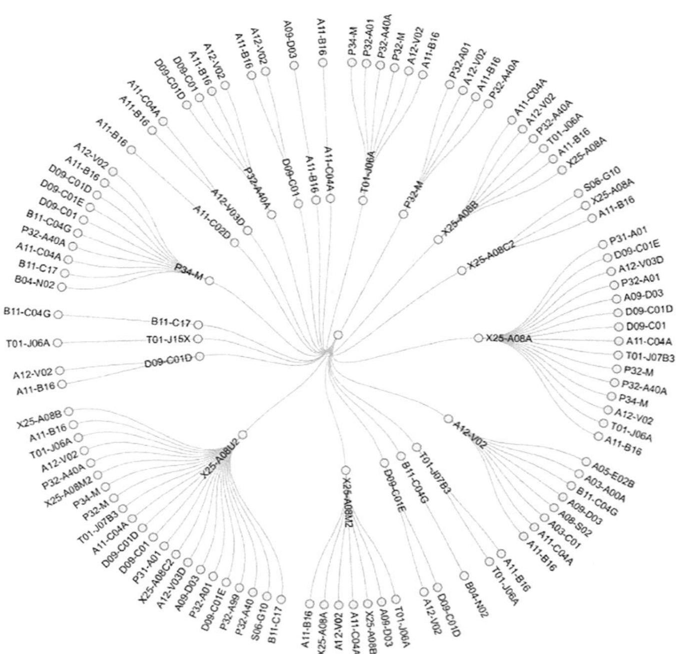

Figure 2.4 Circular dendrogram (manual codes)

The density of the patents comprising a technology domain in the medical, dental, veterinary, and cosmetic and industrial electric equipment classes confirms this information. On the other hand, considering that the amount of patent data obtained at the time of the compilation of the data subjected to the analysis was limited, registration studies for bioprinting technologies will likely increase and expand in terms of technology classes in the future. In this respect, it should be possible to develop a data-based policy and a strategy in decision-making processes for emerging technology domains using analyses similar to those conducted here.

Table 2.3 Bioprinting patents and numbers of registered jurisdictions

TI	No.
Manufacturing of patient-adapted femoral implant for knee joint; involves aligning design for implant relative to substrate in manufacturing apparatus, providing support structures for supporting portions of implant, and detaching structures	21
New hydrophobic peptide and/or peptidomimetic for forming a hydrogel used for producing a wound-dressing or wound-healing agent for the treatment of a wound and a 2D or 3D cell culture substrate	16
Producing gene-activated custom implant used to treat atrophy of bone defects; comprises computed tomography field for bone grafting, simulations based on tomography data, and three-dimensional printing for making of biocompatible carrier medium	14
Method for producing a modification plan for a bespoke guide to orient the glenoid component of a prosthesis within the glenoid cavity of a scapula in shoulder arthroplasty procedures for patients; involves modifying the impression element using a plan	14
Biomaterial with enhanced rubber properties, obtained by naturally cross-linking collagen with hyaluronic acid, useful as, e.g. wound-dressing material, coating material, bone-grafting material, dental material, and food material	14
Method for providing automated 3D printing workflow through a network; involves receiving a request to convert 3D segmented medical image data to print-ready information for printing on three-dimensional printer	13
Composition used to form, e.g. artificial teeth; comprises light-curable viscous mixture composed of, e.g. poly (methyl methacrylate)/methyl methacrylate, multifunctional aliphatic (meth)acrylate, and aliphatic urethane (meth)acrylate oligomer	13
Method of forming images of patterns of color features in displays, e.g. LCDs; involves controlling imaging head to form pixels and different scan-lines of pixels	13
Producing 3D tissue culture model for use in cell assays; involves printing drop of bio-ink to substrate, printing drop of activator to drop of bio-ink to form hydrogel droplet, and repeating printing steps to form a hydrogel mold	12
Forming composition-controlled product using 3D printing for producing dental and optical devices; involves disposing liquid reactant compositions in reservoirs, mixing them and depositing them onto a substrate	10
Bio-ink composition used for bio-printing; comprises bio-blocks composed of a core containing biodegradable polymeric core material and cells and a shell comprising biodegradable polymeric shell material	10
Manufacture of graft scaffold for cartilage repair; involves mixing aqueous solution of gelling polysaccharide, particles, fibers, and/or mammalian cells to obtain a printing mix and depositing the printing mix in a 3D form	10
Luminescent coating used in a dental tool; comprises up-conversion phosphors which emit radiation at a predetermined wavelength matching the initiator activation wavelength of a photoinitiator in a corresponding dental cement	10

THREE-DIMENSIONAL SCANNING TECHNOLOGIES IN MEDICINE: FROM GEOMETRIC SCANNING TO ACCURATELY CAPTURING PART OF AN IMAGE IN A THREE-DIMENSIONAL FORMAT

One of the options in the search for a solution to the constraints of traditional imaging technologies in medicine is 3D scanning technology. Examining the literature, it can be seen that there has been a significant increase in the number of studies on this technology. In this section, commercial ideas developed for use with 3D scanning technologies in medicine are examined through registered patent data. Using data presented on Lens (www.lens.org) as a dataset, patent ownership and the codes with which patents are frequently classified were investigated with a bibliometric analysis. Thanks to peer analysis, the relationships between patent classes could be visualized, and patent citation analysis was used to determine the most effective patent owners and the most cited patent families.

Introduction

This section examines the trend in 3D scanning medical technologies, identifying the companies that dominate the related technology domains and including a case analysis that can be used by technology and engineering managers in related sectors to determine their research focus. Examining the literature, it can be seen that patent analysis is frequently used for TM. Focusing on 3D scanning, it is possible to come across studies examining how 3D scanning technologies are handled in the medical technology domain. (Kellner et al., 2016; Koito et al., 1996; Morimoto et al., 1997). The high commercialization capacity of this technology has attracted the attention of researchers from all research domains, and there has been a leap in the number of publications in the literature (Farhan et al., 2021). Returning to our research focus, it will be useful to provide information on how patent analysis is carried out in order to be able to closely examine the development of 3D imaging/scanning technologies in the medical domain.

Data and Analysis Procedure

As the dataset, we used the bibliographic data of the patents indexed on Lens. The coverage of the Lens database includes information for patents registered in all major patent offices of the world, and the information for patents granted for 3D scanning technologies in the medical domain was analyzed. It is possible to summarize the data sources and the scope of the information from Lens that was used in the dataset as follows:

- The European Patent Office (EPO)'s DOCDB bibliographic data, dating from 1907: more than 81 million documents from nearly 100 jurisdictions;
- United States Patent and Trademark Office (USPTO) applications, dating from 2001, including full texts and images;
- USPTO grants, dating from 1976, including full texts and images;
- USPTO assignments (more than 14 million);
- EPO grants, dating from 1980, including full texts and images;
- World Intellectual Property Organization (WIPO) Patent Cooperation Treaty (PCT) applications, dating from 1978, including full texts and images;
- IP Australia, including full texts.

It is possible to summarize the types of information associated with the patent data provided by Lens as follows: jurisdiction, kind, publication number, Lens ID, publication date, publication year, application number, application date, priority numbers, earliest priority date, title, applicants, inventors, owners (US), URL, type, inclusion of full text, cited by patent count, simple family size, extended family size, sequence count, Cooperative Patent Classification (CPC), Reformed International Patent Classification (IPCR), US classifications, non-patent literature (NPL) citation count, NPL resolved citation count, NPL resolved Lens ID, NPL resolved external ID, and NPL citations.

From examining the literature, we know that 3D scanning technologies have mostly been used for the imaging of the internal organs of living beings (Armanious et al., 2021). Among these technologies, the most common are magnetic resonance imaging (MRI),[1] X-rays,[2] computed tomography (CT),[3] and ultrasonography.[4] On the other hand, since the imaging capabilities of these technologies are not at the same level when imaging the outer parts of living bodies compared to imaging internal organs, R&D activities relating to medical technologies have started to focus on solutions that would enable various limbs to be screened/scanned, especially for living beings with disabilities (Bai & Liu, 2016; Garechana et al., 2019; Jung et al., 2016; Khalil et al., 2016; Rodriguez-Salvador et al., 2017; Tachi et al., 2011). In other words, since the four main imaging technologies mentioned use geometrical imaging that remains limited to the internal organs of the body, it is possible to talk about there being a growing search for a brand new technology for the 3D scanning process to display the outer body (Bo et al., 2011; Gapinski et al., 2014). Here, a trend analysis of the research related to these technology domains was undertaken. In the literature review, it was observed that 3D scanning technologies are handled together with 3D printing technologies

(Haleem & Javaid, 2019). The bibliographic data on the patents in which both technologies are handled together were extracted by selecting only documents containing the subject '3D scanning'. In order to prepare the obtained data for analysis, they had to be processed in a series of cleaning steps and then saved in a relational database.

Results

It was observed that a significant part of the patents for 3D imaging technologies are registered in the US (Figure 2.5). It can be noted that innovation studies of 3D scanning technologies have shown an increasing trend since 2000, and the registration process reached its highest point in 2019. On the other hand, it is also worth noting that, since all the patent data registered until the date of the examination were included, the decline in the number of patents after 2019 was due to the incomplete data from 2020.

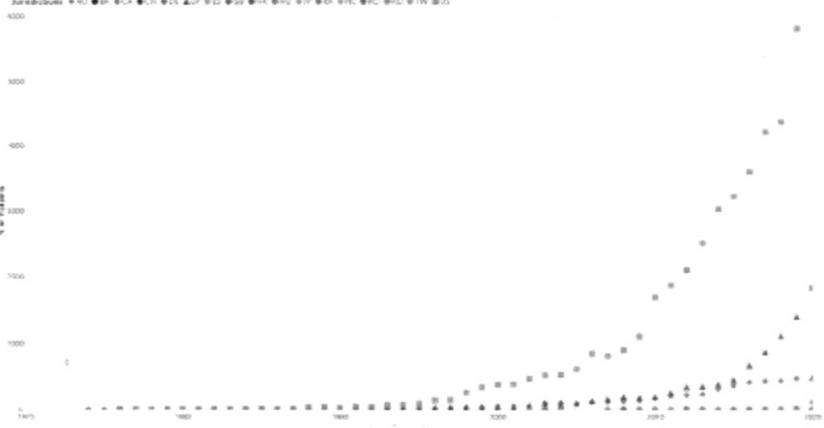

Figure 2.5 Three-dimensional scanning patents by jurisdiction

Both patent ownership and patent analysis data were compiled for the identification of the institutions and companies that dominate 3D scanning technologies. While the numbers of patents were taken to represent ownership when examined alone, we also examined the potential of patents for the analysis of citations from other patents. Accordingly, while the General Electric Company was found to be the company with the highest value in terms of patent value, Siemens Healthcare had the highest value in terms of patent ownership (Table 2.4). In other words, while the General Electric Company has shown an impressive performance in creating a resource for the formation of other

Table 2.4 Three-dimensional scanning patent ownership

No.	Patent owner	h-index	h-core citation sum	All citations	All patents
1	General Electric Company	39	2 932	4 528	229
2	Siemens Medical Solutions USA Incorporated	29	2 596	3 353	167
3	Siemens Corporate Research Incorporated	23	1 318	1 961	102
4	Siemens Aktiengesellschaft	23	1 288	2 119	240
5	Koninklijke Philips Electronics N.V.	21	1 115	1 416	69
6	GE Medical Systems Global Technology Company LLC	20	1 577	1 773	43
7	Siemens Corporation	20	1 159	1 485	108
8	Siemens Healthcare GmbH	19	1 049	1 874	263
9	Kabushiki Kaisha Toshiba	17	1 450	1 651	83
10	Toshiba Medical Systems Corporation	12	687	892	85
11	Microsoft Technology Licensing LLC	11	895	903	19
12	Accuray Incorporated	11	773	777	12
13	Microsoft Corporation	10	883	890	15
14	3M Innovative Properties Company	10	242	310	20
15	Orametrix Incorporated	10	828	830	11
16	Visiongate Incorporated	9	268	268	10
17	Align Technology Incorporated	9	984	992	29
18	Wisconsin Alumni Research Foundation	9	218	242	16
19	GE Medical Systems Global Technology Company LLC	9	421	426	12
20	Koninklijke Philips Electronics N.V.	9	209	374	56
21	Inneroptic Technology Incorporated	9	269	272	14

patents with its innovation ideas, Siemens Healthcare has taken the lead in terms of the number of patents owned.

Through a close examination of the classification codes for 3D scanning technology, it is possible to acquire clues regarding the subdomains of this technology. For this purpose, social network maps were created according to the co-occurrence analysis principles for the CPC codes developed for classification of patents. When the patents were analyzed according to co-occurrence in terms of the classification codes used, they could be grouped in a total of 21 clusters. While the first and largest cluster contained 242 elements, there was only one patent code in the smallest cluster (Figure 2.6).

Accordingly, it was observed that the patents were mostly classified under the class A61B5 ('Detecting, measuring or recording for diagnostic purposes (radiation diagnosis)'). Another important class was the group of patents that

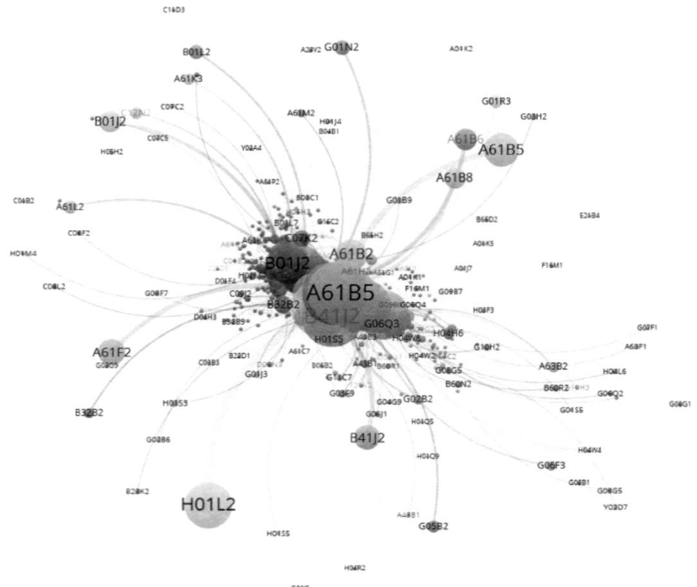

Figure 2.6 CPC code co-occurrence analysis map

were indexed under A61B6 ('Diagnosis by ultrasonic, sonic or infrasonic waves'). It is worth noting that the group B41J2 ('Typewriters or selective printing mechanisms characterized by the printing or marking process for which they are designed') stood out as the third group on the social network map, indicating that 3D scanning technologies are handled together with printing technologies. In this respect, it can be said that the point of view observed in the patenting studies and the point of view in scientific studies are similar. In addition, the H04N1 ('Scanning, transmission or reproduction of documents or the like, e.g. facsimile transmission'), B01J2219 ('Chemical, physical or physico-chemical processes in general'), and G06T7 ('Image analysis') groups were found to be remarkable categories in which the ideas registered for 3D scanning technologies were classified (Table 2.5).

If we look at the status of patents as sources of the production of other patents through an analysis of the technology codes they are related to, it is possible to obtain a performance indicator regarding which patent classes get cited more in other patents. In this way, inferences about which patent families facilitate the production of other patents for a relevant technology can be turned into important indicators that can be used in technology management. For this purpose, the numbers of citations received by each patent from other

Id	Label	Degree	Weighted Degree	Eccentricity	Closeness centrality	Harmonic closeness centrality	Betweenness centrality	Betweenness centrality^2	Bridging coefficient	Bridging centrality	Authority	Hub	Modularity_class	Clustering coefficient	Pageranks	Component number	Cluster number	Cluster ID	Cluster	Clustering	Triangles	Eigencentrality	Dimension_1	Dimension_2
1	A61B5	1	54 133	4	0.441	0.459052	0.000	0.000	85.000	0.000	0.004	0.004	5	0.000	0.001	0	0	12	0	0.000	0.000	0.020	0.044	0.101
2	A61B5	85	151 754	3	0.784	0.864943	0.055	0.055	0.004	0.000	0.174	0.174	5	0.502	0.018	0	1	71	0	0.502	1791.000	1.000	−0.002	0.073
3	A61B6	1	23 463	4	0.408	0.432471	0.000	0.000	69.000	0.000	0.003	0.003	2	0.000	0.001	0	2	9	0	0.000	0.000	0.017	0.805	1.864
4	A61B6	69	76 789	3	0.686	0.788793	0.033	0.033	0.006	0.000	0.152	0.152	2	0.582	0.015	0	3	71	0	0.582	1366.000	0.872	0.291	0.855
5	B41J2	18	108 899	3	0.481	0.533046	0.000	0.000	0.123	0.000	0.030	0.030	0	0.733	0.004	0	4	25	0	0.757	103.000	0.179	0.186	−0.518
6	H04N1	41	74 326	3	0.560	0.647989	0.003	0.003	0.027	0.000	0.088	0.088	0	0.722	0.008	0	5	67	0	0.726	566.000	0.513	1.171	−0.122
7	A61B8	1	21 131	4	0.419	0.441092	0.000	0.000	74.000	0.000	0.003	0.003	1	0.000	0.001	0	6	13	0	0.000	0.000	0.019	0.749	1.145
8	A61B8	74	76 201	3	0.716	0.813218	0.035	0.035	0.005	0.000	0.162	0.162	1	0.569	0.016	0	7	71	0	0.569	1536.000	0.927	0.276	0.454
9	B01J2219	1	20 435	4	0.365	0.390086	0.000	0.000	44.000	0.000	0.002	0.002	7	0.000	0.001	0	8	7	0	0.000	0.000	0.010	−2.393	−0.482
10	B01J2219	44	37 729	3	0.571	0.668103	0.024	0.024	0.011	0.000	0.081	0.081	7	0.551	0.011	0	9	50	0	0.551	521.000	0.472	−1.287	−0.238
11	G06T7	65	52 149	3	0.663	0.762931	0.009	0.009	0.012	0.000	0.147	0.147	2	0.659	0.013	0	10	71	0	0.639	1288.000	0.853	0.654	0.435
12	G06T2207	64	81 580	3	0.655	0.757184	0.008	0.008	0.012	0.000	0.145	0.145	2	0.665	0.013	0	11	71	0	0.644	1258.000	0.842	0.641	0.698
13	H04N2201	27	49 912	3	0.507	0.576149	0.001	0.001	0.058	0.000	0.053	0.053	0	0.793	0.006	0	12	40	0	0.791	257.000	0.312	1.142	0.268
14	H01L2924	1	15 452	4	0.371	0.394397	0.000	0.000	45.000	0.000	0.002	0.002	3	0.000	0.001	0	13	3	0	0.000	0.000	0.011	−0.709	−1.856
15	H01L2924	45	56 380	3	0.586	0.678161	0.026	0.026	0.011	0.000	0.094	0.094	3	0.548	0.011	0	14	65	0	0.548	543.000	0.539	−0.339	−0.876
16	H04N21	1	14 632	4	0.345	0.365661	0.000	0.000	28.000	0.000	0.001	0.001	4	0.000	0.002	0	15	1	0	0.000	0.000	0.007	2.267	1.423
17	H04N21	28	25 497	3	0.525	0.596264	0.018	0.018	0.023	0.000	0.062	0.062	4	0.825	0.008	0	16	42	0	0.825	312.000	0.356	1.237	0.804
18	H01L27	31	36 790	3	0.532	0.603448	0.002	0.002	0.051	0.000	0.069	0.069	3	0.704	0.007	0	17	57	0	0.708	308.000	0.402	0.287	−0.658
19	G01R33	1	10 014	4	0.396	0.421695	0.000	0.000	63.000	0.000	0.003	0.003	5	0.000	0.001	0	18	2	0	0.000	0.000	0.016	−0.506	2.090
20	G01R33	63	23 426	3	0.652	0.758621	0.026	0.026	0.007	0.000	0.144	0.144	5	0.628	0.014	0	19	71	0	0.628	1227.000	0.826	−0.270	1.007

Table 2.5 *Network statistics for CPC code co-occurrence analysis map*

patents were compiled and then the total numbers of patent citations received by each patent classification code were calculated. After the process, by taking the intersection of the numbers of citations of each classification code and the numbers of patents, the h-index values were calculated for the patent codes (Table 2.6).

Table 2.6 CPC citation analysis

No.	Patent classification	h-index	h-core citation sum	All patent citations	All patents
1	A61B5	250	103 186	416 238	14 803
2	A61B2034	234	99 291	170 641	2 860
3	A61B2090	234	92 171	160 804	3 296
4	A61B8	205	61 648	181 920	8 313
5	A61B2017	196	69 438	124 384	2 548
6	A61B34	183	68 372	105 082	1 979
7	A61B6	180	53 419	187 495	10 302
8	A61B90	175	60 640	94 993	1 643
9	A61B17	168	53 946	109 295	2 835
10	G05B2219	150	59 394	83 360	880
11	B01J2219	143	37 735	77 961	2 062
12	H04N13	139	53 635	74 014	1 094
13	G06T2207	136	35 578	102 109	8 699
14	H04N21	133	60 902	84 498	1 242
15	H01L2924	131	31 516	59 158	1 732
16	A61B2018	129	40 306	66 911	2 324
17	G01S7	125	25 932	55 543	1 938
18	G01S15	124	27 282	51 700	1 482
19	A61F2002	122	25 432	46 828	1 668
20	A61F2	119	24 970	49 546	1 791

Discussion

As a result of the analyses made, it was observed that both the results obtained for scientific articles and the results we obtained from the patent analysis were similar. In particular, the handling of 3D scanning technologies together with 3D printer technologies was among the most important similarities obtained in this context. However, by emphasizing the importance of patents citing other patents as a patent ownership indicator in the patent analysis, we were also able to see which other commercial ideas the original information obtained constitutes the basis for. We should say that the biggest deficiency in patent classifications is that they do not provide details that allow for sector analysis. However, structural gap analysis has significant potential for sectoral analyses of patents. In this way, in addition to the visualization of a technology network, different evaluations and different inferences with text-mining algorithms can contribute to technology and engineering management. More realistic results can be obtained in investment decisions with the use of analyses carried out by taking into account the speed of technology diffusion. These results can provide very critical information for managers who are in decision-making positions in engineering and technology management.

WIRELESS POWER TECHNOLOGY: SEEKING THE HERITAGE OF TESLA

Today, with the development of technologies, it can be seen that the need for energy is increasing. It is a fact that this increasing energy need is forcing users to connect to a certain network. Since wireless charging technology, the inductive technology that has been developed to remove the obstacles to the independence of the end user, is a technology that is more and more present in our lives, the increase in the number of studies on the subject at both the basic and product research levels stands out. For example, as a solution to the problems experienced by end users, wireless charging technologies basically produce devices at the level of consumer electronics at three main power values: wireless chargers of 5 watts for wearable technologies, 15 watts for fast charging and smartphone charging, and 30 watts for laptops and tablets. In this part of the study, wireless charging technology was analyzed based on patent data.

Wired solutions were used in traditional methods for electricity transfer. However, today's technological progress has led to the search for wireless technologies, especially in the charging of mobile devices. Wireless power transmission refers to the transmission of electrical energy through electric, magnetic, or electromagnetic fields instead of wires or cables (Brown, 1996). Wireless power transmission can be used in situations interconnected cables would be dangerous or impossible to use (Sumi et al., 2018). Wireless power transmission is widely used in electric toothbrush chargers, implanted medical devices (such as artificial pacemakers), RFID tags, and many other applications. The basic concepts for wireless communication and wireless power transmission are similar in that they use the same fields and waves, but the main goal in wireless communication is to transfer data. In wireless communication, the transmitted power is not important, as long as the signal-to-noise ratio of the received signal is acceptable. Wireless power transmission aims to maximize power transmission efficiency and/or coverage while ensuring the health and safety of the general public (Zeng et al., 2017).

Patents provide important outputs for technology intelligence. When the literature is examined, analyses of wireless power are frequently encountered (An et al., 2018; Bates, 2020; Jeong & Kim, 2014; Manohar et al., 2018; Naumanen et al., 2019; Saritas & Burmaoglu, 2016). In this part of the study, the development, evolution, and prominent areas of wireless power technologies along with their sub-technology domains were determined. As in the previous sections, bibliometric and network analysis methods were used. Data from Lens were used as the dataset. Only the 'granted patent' data were examined for the document types analyzed. The total number of patents subjected to analysis was 18 160 (Figure 2.7).

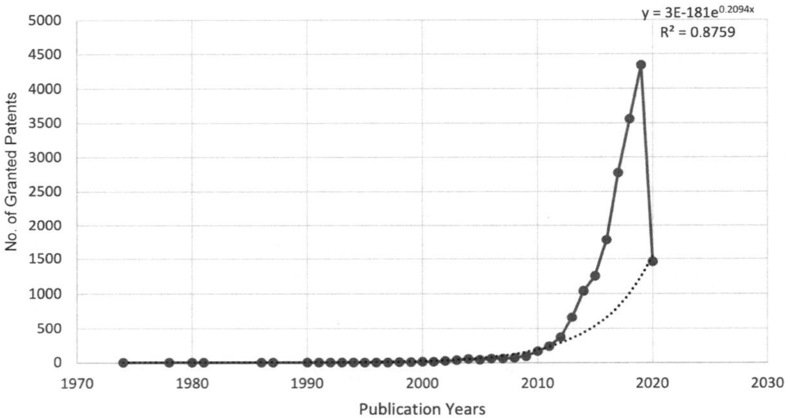

Figure 2.7 Number of patents granted for wireless power

With regard to the annual growth rate, it was observed that the number of patents registered for wireless power technologies corresponded to an exponential growth curve. The R value also indicated that the correlation level of descriptors was quite high. In this respect, it is possible to say that patents for wireless power technologies show a geometric growth trend. It should also be noted that the decrease in the number of patents for 2020 was due to the incomplete indexing of the documents of the relevant year at the time of the patent data analysis. Analyzing patent ownership gives us the opportunity to determine the institutions and organizations that direct the technology. When this was examined by adding attribution values to this indicator, it was possible to identify the institutions that provide infrastructure for new ideas. In this regard, the patent holders, the number of patents, and the total citations received were determined. Also added to these metrics were the h-index values which represent the intersection points of patent owners' citation numbers and patent numbers (Table 2.7).

Table 2.7 Patent owners and citation metrics

Patent owner	h-index	h-core citation sum	All citations	All patents
Witricity Corporation	21	2 037	2 218	96
Qualcomm Incorporated	18	1 455	1 831	130
Energous Corporation	18	1 549	1 775	65
Samsung Electronics Company Ltd	13	792	1 067	166
Philips IP Ventures B.V.	11	344	395	36
Access Business Group International LLC	9	299	338	20
LG Electronics Incorporated	8	226	286	48
TDK Corporation	7	266	314	39
Nigel Power LLC	6	100	107	10
Panasonic Corporation	6	232	245	13
Semiconductor Energy Laboratory Company Ltd	6	315	337	15
Koninklijke Philips N.V.	6	84	113	27
Avago Technologies General IP	5	279	289	24
Avago Technologies International Sales Pte Ltd	5	279	289	24
Panasonic Intellectual Property Management Company Ltd	5	227	233	13
Hanrim Postech Company Ltd	5	79	119	40
GE Hybrid Technologies LLC	5	34	68	38
LG Innotek Company Ltd	5	220	284	69
Wits Company Ltd	5	228	259	48
Intel Corporation	4	243	247	20

Looking at patent applicants, it can be seen that institutional applications come to the forefront (Table 2.8). While it can be observed that patent applicants are generally technology companies, which are considered as legal entities, there are also representatives from sectors such as information and communication technologies, the automotive industry, and banking. The Louvain community detection algorithm was used to determine the number of clusters that appeared in the SNA. In this way, it was possible to group the dominant emerging research domains.

In the SNA undertaken on the class codes of patents, nine clusters were identified.[5] Accordingly, there were 121 elements in the cluster that ranked first in terms of network values (Figure 2.8).

In terms of betweenness, one of the patents registered in the field of wireless power was for 'Charging batteries from ac mains by converters', 'Arrangements for balancing the load in a network by storage of energy', 'Near-field transmission systems, e.g. inductive loop type', 'Using inductive coupling', and 'Measuring for diagnostic purposes'. It was observed that patents in the 'Identification of persons' classes were at the top of the list. This shows that wireless power transfer technologies are primarily related to data transfer. In this respect, wireless power transfer should be considered as a very important component for the connected device ecosystem, especially the Internet of Things (IoT).

Table 2.8 Patent applicants for wireless power technologies

Patent applicant	h-index	h-core citation sum	All citations	All patents
AT&T Intellectual Property	86	17 839	19 049	414
Automotive Technologies International	39	5 815	5 978	48
Witricity Corporation	36	3 776	4 492	237
Karalis, Aristeidis	36	3 753	4 177	54
Soljacic, Marin	35	3 567	3 910	51
Kesler, Morris P.	33	3 148	3 397	53
Hall, Katherine L.	33	3 143	3 359	52
Kurs, Andre B.	31	2 627	2 750	42
Access Business Group International LLC	30	2 490	3 223	164
iRobot Corporation	26	1 122	1 683	82
Qualcomm Incorporated	25	2 861	4 270	560
Campanella, Andrew J.	25	2 162	2 216	30
Baarman, David W.	24	929	1 239	53
Massachusetts Institute of Technology	23	3 105	3 196	61
Kulikowski, Konrad J.	23	2 132	2 145	26
Whirlpool Corporation	21	670	857	42
Energous Corporation	21	2 695	3 915	213
Joannopoulos, John D.	19	1 874	1 885	21
Mojo Mobility Incorporated	19	2 130	2 161	29
Shell Oil Company	17	762	803	21

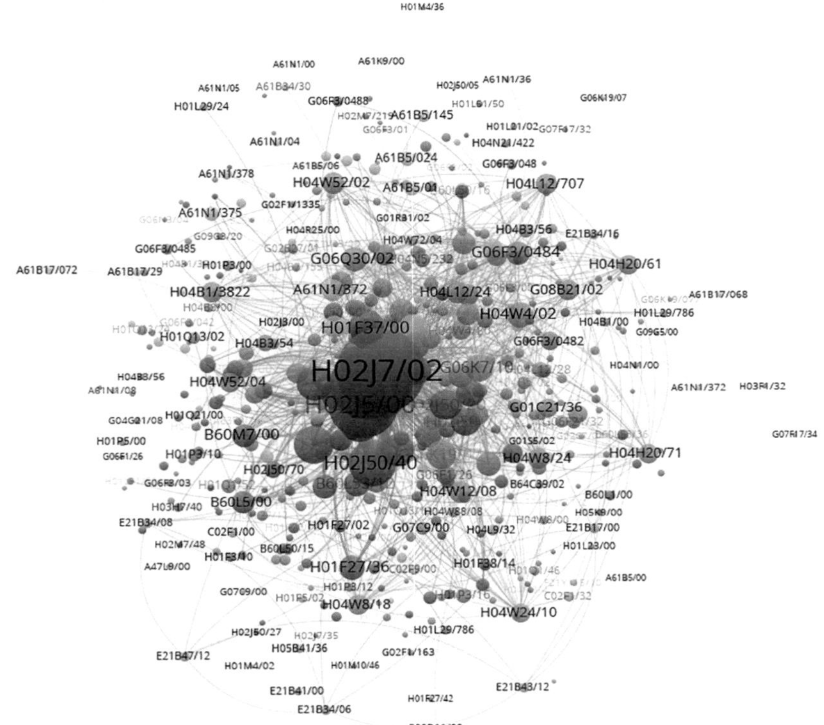

Figure 2.8 *International Patent Classification code network for wireless power technology*

UNMANNED AGRICULTURE: ANALYSIS OF THE USE OF MECHATRONIC TOOLS IN AGRICULTURAL ACTIVITIES

Agriculture can be defined as all the maintenance, feeding, breeding, protection, and mechanization activities relating to all the different kinds of plant/animal products that can be used as human food and have economic value, as well as all the fishing activities carried out in stagnant waters or in private areas. Agricultural activities, which have been affected by many developments in the historical process, display a pattern of frequently benefitting from technology. In this section, commercialization activities for the agricultural use of drone and robotic technologies, which have a wide range of applications, from defense technologies to production technologies, are examined. For this

Number	International Patent Classification	Patent classification	Multi-level Louvain communities in N3*	All-degree partition	Weak components	All neighbors of vertex	All core partitions	Hubs and authorities	Size of all domains	All degrees	All weighted degrees	All closeness centralities	Betweenness centrality	Hub weights	Authority weights	Average distance from all domains	All proximity prestige values	Aggregate constraint
2	H02J7/02	For charging batteries from ac mains by converters	1	514	1	1	50	2	499	514	15 381	0.809	0.073	0.488	0.488	1.236	0.809	0.103
1	H02J7/00	Arrangements for balancing the load in a network by storage of energy	1	511	1	0	50	2	499	511	10 952	0.771	0.073	0.366	0.366	1.297	0.771	0.112
18	H04B5/00	Near-field transmission systems, e.g. inductive loop type	2	439	1	1	50	0	499	439	3841	0.733	0.064	0.105	0.105	1.365	0.733	0.076
7	H02J50/10	Using inductive coupling	1	439	1	1	50	2	499	439	6014	0.749	0.049	0.220	0.220	1.335	0.749	0.127
3	H02J5/00	Circuit arrangements for transfer of electric power between ac networks and dc networks	1	411	1	1	50	2	499	411	9448	0.730	0.041	0.346	0.346	1.371	0.730	0.128
49	A61B5/00	Measuring for diagnostic purposes; identification of persons	4	276	1	1	50	0	499	276	1738	0.635	0.036	0.010	0.010	1.575	0.635	0.049
4	H02J50/12	Of the resonant type	1	402	1	1	50	2	499	402	9227	0.716	0.034	0.344	0.344	1.397	0.716	0.133
46	G06F1/16	Constructional details or arrangements	3	277	1	1	50	0	499	277	1752	0.633	0.027	0.014	0.014	1.579	0.633	0.055
6	H01F38/14	Inductive couplings	1	342	1	1	50	2	499	342	7358	0.668	0.024	0.246	0.246	1.497	0.668	0.115
5	H04B5/00	Near-field transmission systems, e.g. inductive loop type	1	218	1	1	50	2	499	218	5605	0.639	0.018	0.229	0.229	1.565	0.639	0.137
8	H02J50/80	Involving the exchange of data, concerning supply or distribution of electric power, between transmitting devices and receiving devices	1	313	1	1	50	2	499	313	5405	0.677	0.017	0.219	0.219	1.477	0.677	0.147
23	H02J50/20	Using microwaves or radio frequency waves	1	283	1	1	50	0	499	283	2043	0.653	0.016	0.068	0.068	1.531	0.653	0.113
37	G06F3/041	Digitisers, e.g. for touch screens or touch pads, characterised by the transducing means	3	188	1	1	50	0	499	188	1248	0.581	0.013	0.003	0.003	1.721	0.581	0.088
9	H02J50/40	Using two or more transmitting or receiving devices	1	264	1	1	50	2	499	264	4441	0.643	0.010	0.186	0.186	1.555	0.643	0.157
17	B60L11/18	Using power supplied from primary cells, secondary cells, or fuel cells	1	233	1	1	50	0	499	233	3233	0.603	0.009	0.114	0.114	1.657	0.603	0.130
11	H01F27/42	Circuits specially adapted for the purpose of modifying, or compensating for, electric characteristics of transformers, reactors or choke coils	1	231	1	1	50	0	499	231	3147	0.624	0.009	0.110	0.110	1.603	0.624	0.147
58	G06F3/01	Input arrangements or combined input and output arrangements for interaction between user and computer	3	201	1	1	50	0	499	201	1230	0.592	0.008	0.002	0.002	1.689	0.592	0.066
157	G06F19/00	Digital computing or data processing equipment or methods, specially adapted for specific applications	5	150	1	1	50	0	499	150	974	0.573	0.008	0.001	0.001	1.745	0.573	0.064
20	H02J50/90	Involving detection or optimisation of position, e.g. alignment	1	232	1	1	50	0	499	232	2158	0.621	0.008	0.087	0.087	1.611	0.621	0.143
79	H04M1/02	Constructional features of telephone sets	3	172	1	1	50	0	499	172	692	0.580	0.008	0.006	0.006	1.725	0.580	0.063
77	H04M1/725	Cordless telephones	3	212	1	1	50	0	499	212	951	0.593	0.007	0.003	0.003	1.685	0.593	0.056

Note: *N3: 500, res=1.000000, Q=0.458898, NC=9.

Table 2.9 Network parameters for International Patent Classifications

Table 2.10 International Patent Classification citation analysis for wireless power technology

International Patent Classification	h-index	h-core citation sum	All citations	All patents
H02J7/00	49	7 318	12 383	2 697
H02J7/02	45	7 263	11 437	3 261
H02J5/00	34	4 112	8 002	2 065
H04B3/52	31	6 401	6 612	186
H04B5/00	31	3 565	5 735	1 397
H01F27/42	30	3 098	4 981	643
H01F37/00	29	2 934	4 143	446
H01F38/14	28	2 393	4 511	1 592
H02J17/00	28	3 716	4 618	327
H02J7/00	27	3 301	4 158	505
H04B3/54	26	4 253	4 460	158
H02J7/02	26	4 145	5 932	839
B60L11/18	25	1 450	2 974	751
H02J17/00	25	4 087	4 733	423
H01P3/16	24	4 170	4 246	91
H01P3/10	24	3 908	3 964	81
H01F38/00	24	1 808	2 038	56
H04B1/38	23	3 499	3 834	155
G06F19/00	21	1 401	1 527	42
H04B5/00	21	1 590	2 978	877

purpose, the ownership of the ideas registered as patents and their distribution in different countries were examined closely. Furthermore, to determine the main technology areas and sub-technology domains, SNA based on the patent classes was undertaken.

Introduction

Robotic technologies are preferred for the follow-up and execution of regular jobs that humans are not able to carry out or that need to be put into a certain routine program. With the development of mechatronic devices, the flexibility required to incorporate devices of any size at any stage of agriculture was achieved. In this respect, it can be said that drone and robotic technologies are included in agricultural processes both to support the labor force and to automate the operations that need to be done. Based on these developments, this part of the study focused on the analysis of drone and robotic technologies in agricultural processes using patent data. Before moving on to the analysis phase, it would be useful to provide some information and mention some case studies about TM and the use of patent documents. With TM, complex data can be transformed into meaningful and usable information and used to shorten decision-making processes. TM aims to provide a fast, reliable, and healthy decision-making environment (Tseng et al., 2007; Yoon & Park, 2004). Many sub-analysis methods, notably patent analysis, are used within this framework. Patent analysis has been shown to be among the methods frequently used in TM (Daim et al., 2006; Daim et al., 2012; Daim et al., 2020). By using patent analysis, research on the technology ownership of enterprises can be undertaken (Daim et al., 2012; Daim et al., 2020; Maeno et al., 2011) or on the change of a particular technology over time (Karvonen & Kässi, 2011).

Patent documents indexed in international patent offices on the use of robotic technologies and drones in agricultural processes were examined. Usage areas of robotic technologies in agricultural activities were thereby determined. For the detection of the technologies that have been commercialized up to the present, scientometrics, bibliometrics, and SNA techniques were used together and the technology areas that have reached the saturation point were determined with the resulting analysis. However, to determine the level of technological maturity in terms of technology ownership, analyses were conducted on applicant ownership information and a matrix was developed based on technology and applicant information (Chang, 2012). This is an important tool that can be used to determine the status of technological information, identify companies and competitors, and reveal the dominant actors that should be monitored (Daim et al., 2006). In this regard, regardless of the sector, it is possible to obtain significant data for decision making with TM. TM is of critical importance in providing the qualified information needed to eliminate uncertainty in the decision-making process (Li et al., 2009; Tseng et al., 2007; Yalçin et al., 2020; Yu & Lo, 2009). It is possible to state that patent data is used extensively for this purpose.

Data and Analysis

It is known that patents provide important data in the context of commercialization. In this respect, patent ownership, which is closely related to the development level of countries, has begun to be used as a very important indicator of the technologies and sub-technology domains that patents are related to. The patent data, which are examined together with visuals illustrating, for example, the novelty, originality, radicalness, and preliminary and successor attribution statuses of the idea produced, have begun to be used in technology foresight calculations in order to make short-term decisions and in the calculation of long-term decisions. It is therefore possible to state that patent data is used extensively in TM at points such as the creation of competitive advantage and the generation of innovative information. In this context, R&D activities conducted for the use of robots/drones in agricultural applications or the integration of drones into agricultural processes were examined. Lens was used as the data source because it is considered to be a reliable database to use for patent analyses (Castelló-Cogollos et al., 2018; Sydor, 2019). We built the first dataset through a lexical search strategy using the terms '(drones or robot*) and (agricultu*)'. A total of 2945 patent documents that met these criteria were accessed. The obtained data were recorded in a relational database in preparation for analysis. Growth curves, which are among the methods frequently deployed in patent analysis, were used (Choi et al., 2015; Pantano et al., 2017). The annual growth rate of documents related to registered ideas about drone or robotic technologies in agricultural activities fit an exponential growth curve ($R^2 = 0.8733$). This can be interpreted as implying that the patents for drone or robotic technologies in agricultural activities show a pattern in accordance with an exponential growth curve. In other words, the increase in the amount of patents relating to drone or robotic technologies produced for agricultural activities demonstrates geometrical growth (Figure 2.9).

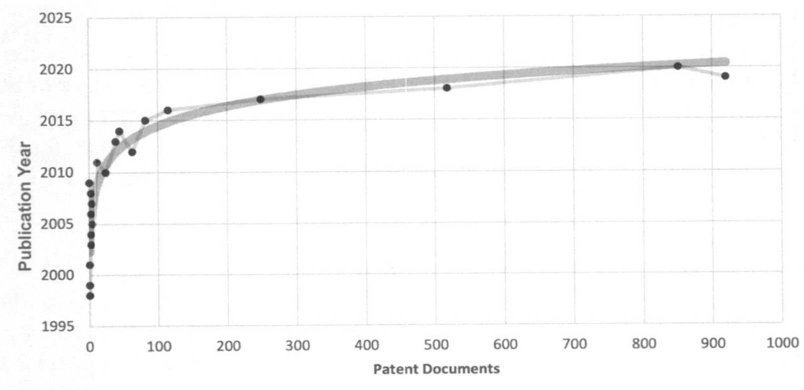

Figure 2.9 Circular dendrogram (manual codes)

Results

SNA was conducted on patent class codes to identify technology areas where technologies for drones/robots intended to be developed for unmanned production or effective production in agricultural activities are concentrated. In this way, the main technology domains were determined. Furthermore, technologies reaching critical density or a level of maturity and subareas open to improvement were determined with the parameters obtained from the SNA. Structural gaps identified in the SNA workflow are among the indicators frequently used for this purpose (Table 2.11). Looking at the first analysis results, it can be observed that the domain of digital computers is the technology domain in which application efforts are most intense. Mechanical engineering and tools is the second most intense technology class for agricultural applications, followed by hand tools and cutting, transport, plant culture and dairy products, and process and machine control.

When we look at the locations of these patents on the map in terms of their centrality values in the network, it is possible to make sense of the roles they play and the technology domains they are concentrated in (Figure 2.10). Considering these results, it is possible to state that the search for computational features in drone and robotic technologies is a priority technology domain. On the other hand, it can be seen that the search for sub-technology domains in which computational processes are handled together with mechatronic parts is a second important research foci. From these two main focal points arising from basic needs, it can be seen that technology domains for agricultural activities find a place in this search. In this regard, it is clear that the search for the integration of the mentioned technology areas into agricultural processes is still continuing (Figure 2.10).

When this visual is examined closely, it can be seen that the W06 class ('Aviation, marine and radar systems') is positioned close to the center of the network. If we evaluate this framework in general, we can note that the patent classes W06, P11, P13, Q25, and P14 are important cluster elements for the drone and robotic technologies network in agricultural processes.

Patent Ownership

Examining the use of drone and robotic technologies in agricultural activities through patent ownership, it can be seen that there are mostly two large clusters in commercialization activities. In addition, it is possible to state that four medium-sized clusters are present (Figure 2.11). In the patent ownership network, in which real- and legal-person clusters appear together, it can be seen that Qualcomm stands out in the first cluster, and Witricity Corporation stands out in terms of network values in the second cluster. Qualcomm's name stands out due to the high demand for mobile processors and small-sized control units, while Witricity Corporation's prominence represents the high demand for consumer electronics and wireless charging technologies. The figure also addresses the modularity and wireless charging needs of drones and

Table 2.11 *Network parameters for mechatronic tools in agricultural activities (patent classifications)*

Label	All-degree partition	Hubs and authorities	All degrees	All weighted degrees	All closeness centralities	Betweenness centrality	Hub weights	Authority weights	Average distance from all domains	All proximity prestige	Aggregate constraint
T01 (Digital computers)	101	2	101	1 306	0.586	0.087	0.524	0.524	1.707	0.586	0.137
A88 (Mechanical engineering and tools e.g. valves, gears and conveyor belts)	95	0	95	196	0.571	0.111	0.027	0.027	1.753	0.571	0.058
P62 (Hand tools, cutting (B25, B26))	92	1	92	507	0.575	0.101	0.243	0.243	1.739	0.575	0.207
A95 (Transport – including vehicle parts, tyres and armaments)	91	0	91	213	0.551	0.063	0.016	0.016	1.815	0.551	0.064
P13 (Plant culture, dairy products (A01G, H, J))	90	2	90	743	0.576	0.090	0.287	0.287	1.735	0.576	0.148
T06 (Process and machine control)	81	2	81	782	0.557	0.039	0.423	0.423	1.794	0.557	0.221
X25 (Industrial electric equipment)	77	2	77	691	0.554	0.077	0.331	0.331	1.805	0.554	0.188
A85 (Electrical applications)	65	0	65	120	0.500	0.021	0.008	0.008	2.000	0.500	0.070
P11 (Soil working, planting (A01B, C))	62	1	62	397	0.537	0.033	0.159	0.159	1.861	0.537	0.152
P12 (Harvesting (A01D, F))	61	0	61	192	0.528	0.025	0.072	0.072	1.895	0.528	0.153
W06 (Aviation, marine and radar systems)	59	0	59	273	0.524	0.017	0.084	0.084	1.909	0.524	0.118

Label	All-degree partition	Hubs and authorities	All degrees	All weighted degrees	All closeness centralities	Betweenness centrality	Hub weights	Authority weights	Average distance from all domains	All proximity prestige	Aggregate constraint
W06 (Aviation, marine and radar systems)	58	2	58	712	0.528	0.032	0.381	0.381	1.895	0.528	0.212
S02 (Engineering instrumentation)	57	0	57	246	0.516	0.021	0.104	0.104	1.937	0.516	0.175
P14 (Animal management and care (A01K, L, M))	56	1	56	283	0.531	0.038	0.114	0.114	1.882	0.531	0.246
A97 (Miscellaneous goods not specified elsewhere – including papermaking, gramophone records, detergents, food and oil well applications)	56	0	56	98	0.517	0.047	0.012	0.012	1.934	0.517	0.091
S03 (Scientific instrumentation)	55	0	55	180	0.522	0.016	0.067	0.067	1.916	0.522	0.136
T01 (Digital computers)	53	0	53	166	0.509	0.045	0.037	0.037	1.965	0.509	0.104
W04 (Audio/video recording and systems)	51	1	51	312	0.507	0.008	0.114	0.114	1.972	0.507	0.149
W01 (Telephone and data transmission systems)	50	0	50	217	0.504	0.009	0.083	0.083	1.986	0.504	0.167
A89 (Photographic, laboratory equipment, optical – including electrophotographic, thermographic uses)	50	0	50	88	0.507	0.016	0.006	0.006	1.972	0.507	0.091

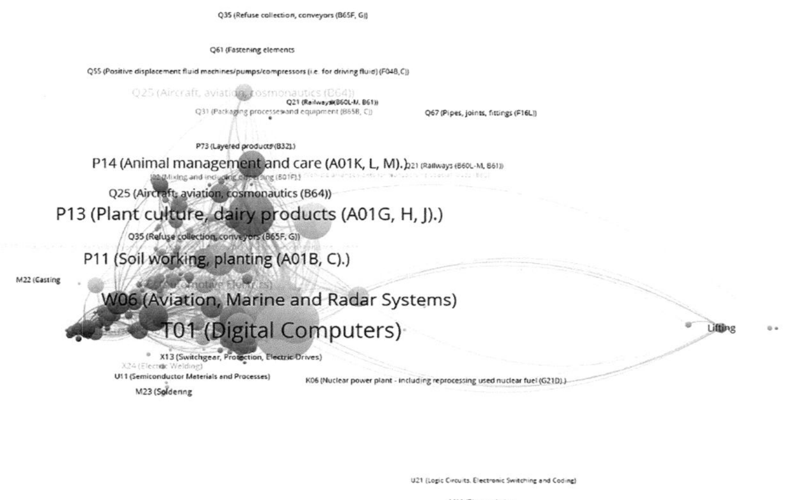

Figure 2.10 *Network map for mechatronic tools in agricultural activities (degree)*

robotic devices that are intended to be used in agricultural activities, as well as providing clues about the search for solutions.

It is possible to list the most efficient network elements in patent ownership as follows: Kesler, Morris P.; Qualcomm; Kurs, Andre B.; Karalis, Aristeidis; Hall, Katherine L.; Campanella, Andrew J.; Witricity Corporation; Kulikowski, Konrad; Soljacic, Marin; and Fiorello, Ron. The names other than Qualcomm and Witricity are those of real people. However, it can be noted that Morris Kessler is working as CTO of Witricity (Table 2.12). It is possible to state that these applicants and patent holders are the dominant actors with regard to research on drone and robotic technologies in agricultural activities. In this respect, the technological knowledge status of these applicants can be understood to have reached a level of maturity. It would be beneficial to consider this table in terms of the investments to be made in the relevant technology. Furthermore, the calculation of the network values (aggregate constraints) and the determination of structural gaps would also allow us to identify the network elements that are either developing or have not yet reached a critical point for specialization.

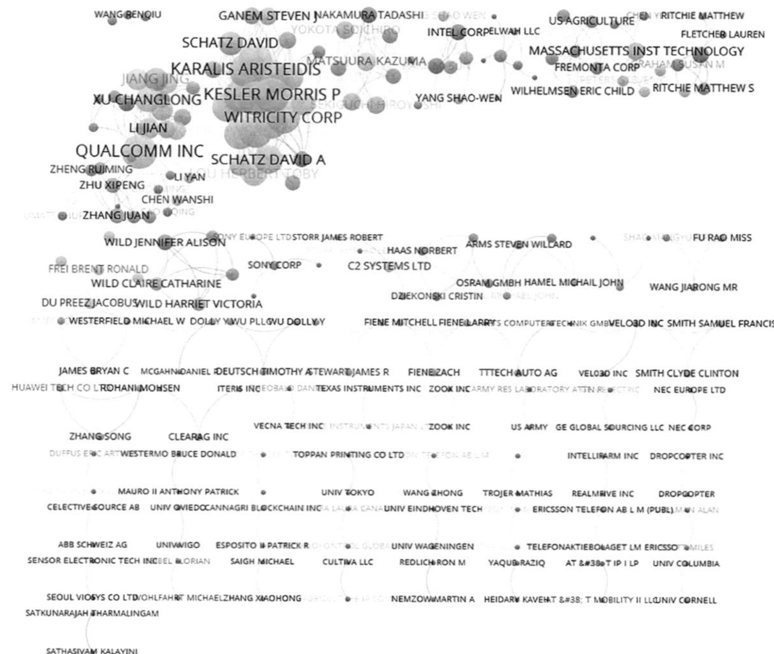

Figure 2.11 *Assignee network for mechatronic tools in agricultural activities (degree)*

Discussion

It has been observed that technology ownership, dominant technologies, and sub-technology domains could be determined with the results obtained. The metrics and measures related to technology density and maturity level stand out both in the strategic decision-making process and as instruments that can create a competitive advantage. Identifying technologies that create critical density can provide benefits that can be used in directing R&D investments correctly and defining the competitive ecosystem. However, the detection of structural gaps is critical to identify technologies that are open to improvement. Identifying converging and diverging technologies can be listed among the options offered by TM as a guiding tool for investment in the right technology. It should not be forgotten that documents intended for registered ideas are used in patent analysis. The most important limitation of the method was that the dataset analyzed did not include the same scope for all countries. The data

Assignee	All-degree partition	Periodic strong components	Multi-level Louvain communities	Hubs and authorities	All degrees	All weights	All closeness centralities	Betweenness centrality	Hub weights	Authority weights	Normalized size of all domains	Average distance from all domains	All proximity prestige values	Aggregate constraint
Kesler, Morris P.	37	1	5	2	37	771	0.108	0.001	0.411	0.411	0.104	1.000	0.104	0.321
Qualcomm	36	2	14	0	36	123	0.132	0.015	0.000	0.000	0.149	1.152	0.129	0.166
Kurs, Andre B.	35	1	1	2	35	698	0.099	0.000	0.389	0.389	0.104	1.087	0.095	0.339
Karalis, Aristeidis	35	1	3	2	35	621	0.095	0.000	0.357	0.357	0.104	1.130	0.092	0.353
Hall, Katherine L.	35	1	4	2	35	759	0.108	0.001	0.408	0.408	0.104	1.000	0.104	0.324
Campanella, Andrew J.	34	1	6	2	34	543	0.092	0.000	0.316	0.316	0.104	1.174	0.088	0.359
Witricity Corporation	33	1	7	2	33	335	0.099	0.000	0.190	0.190	0.104	1.087	0.095	0.350
Kulikowski, Konrad	32	1	10	2	32	278	0.095	0.000	0.151	0.151	0.104	1.130	0.092	0.333
Soljacic, Marin	31	1	2	2	31	613	0.092	0.000	0.355	0.355	0.104	1.174	0.088	0.356
Fiorello, Ron	31	1	12	0	31	209	0.099	0.000	0.100	0.100	0.104	1.087	0.095	0.303
Giler, Eric R.	28	1	9	2	28	302	0.085	0.000	0.160	0.160	0.104	1.261	0.082	0.328
Schatz, David A.	26	1	11	0	26	240	0.083	0.000	0.129	0.129	0.104	1.304	0.079	0.333
Li, Qiang	26	1	13	0	26	163	0.083	0.000	0.090	0.090	0.104	1.304	0.079	0.341
Kulikowski, Konrad J.	25	1	8	2	25	323	0.083	0.000	0.193	0.193	0.104	1.304	0.079	0.369
Xu, Changlong	18	2	15	0	18	80	0.099	0.001	0.000	0.000	0.149	1.545	0.096	0.316
Efe, Volkan	18	1	16	0	18	61	0.073	0.000	0.035	0.035	0.104	1.478	0.070	0.363
Lou, Herbert Toby	18	1	23	0	18	48	0.073	0.000	0.025	0.025	0.104	1.478	0.070	0.328
Jiang, Jing	17	2	15	0	17	44	0.093	0.000	0.000	0.000	0.149	1.636	0.091	0.352
Schatz, David	17	1	24	0	17	43	0.073	0.000	0.024	0.024	0.104	1.478	0.070	0.353
Wu, Liangming	15	2	15	0	15	36	0.093	0.000	0.000	0.000	0.149	1.636	0.091	0.360

Table 2.12 Network parameters of the assignee network for mechatronic tools in agricultural activities (degree)

and results obtained should be interpreted considering these constraints, and the decisions to be taken should be evaluated within this framework.

NOTES

1. 'Magnetic resonance imaging (MRI) is a medical imaging technique used in radiology to form pictures of the anatomy and the physiological processes of the body. MRI scanners use strong magnetic fields, magnetic field gradients, and radio waves to generate images of the organs in the body. MRI does not involve X-rays or the use of ionizing radiation, which distinguishes it from CT and PET scans. MRI is a medical application of nuclear magnetic resonance (NMR). NMR can also be used for imaging in other NMR applications, such as NMR spectroscopy' (Wikipedia, 2020a).
2. 'X-rays are a type of radiation called electromagnetic waves. X-ray imaging creates pictures of the inside of your body' (MedlinePlus, 2020).
3. 'The term "computed tomography", or CT, refers to a computerized x-ray imaging procedure in which a narrow beam of x-rays is aimed at a patient and quickly rotated around the body, producing signals that are processed by the machine's computer to generate cross-sectional images – or "slices" – of the body' (National Institute of Biomedical Imaging and Bioengineering, 2020).
4. 'Medical ultrasound (also known as diagnostic sonography or ultrasonography) is a diagnostic imaging technique, or therapeutic application of ultrasound. It is used to create an image of internal body structures such as tendons, muscles, joints, blood vessels, and internal organs' (Wikipedia, 2020b).
5. Number of clusters: 9, modularity: 0.458898, Louvain community detection, coarsening and refinement: 4. Multi-level Louvain communities in N3: 500, res=1.000000, Q=0.458898, NC=9.

3. Network-based analyses

The creation of technology networks is among the methods frequently used in the study of the development of technologies. It is possible to create a technology network for a technology using patent data. In this way, it is possible to determine the phase of the examined technology in its life cycle (S-curves, etc.) and to map the technology networks based on the IPC codes to which patents belong. A strategic technology code network connected to IPC codes can be created with the visualization of the technology networks that have been created. The directions of technologies, that is, the relationships that nodes establish with each other, may differ. In some technology networks, certain technologies can only establish one-way relationships with others, whereas other relationships can also be bidirectional or simultaneously non-directional. An actor's direct relationship with other actors in the network is called the degree. The number of direct connections to an actor from other actors is called the in-degree, the number of direct connections that an actor sends to other actors is called the out-degree. The extent of one actor's reach to another is called the closeness, and the value expressing the level of actors' closeness to each other is called the density. Structural gaps are among the network metrics that enable individuals or actors to connect to different network groups via other network actors. In TM and SNA, they are important for the detection of critical node points in the network (Burt, 1995; Crona et al., 2011; Ehrlich & Carboni, 2005). In this part of the study, details about the metrics and indicators of SNA are given.

UTILIZING SOCIAL NETWORK ANALYSIS IN TECHNOLOGY MANAGEMENT: IDENTIFYING KEY ACTORS FOR AUTOMATED VEHICLES

The average number of sectoral technology structures owned by institutions is dependent on the competitiveness of countries. The ability of companies to integrate their technology strategies into international dynamics is directly related to the ability of countries to produce technology or technology transfers. Technology production depends on public and private sector co-operation and the quality and competencies of the intellectual capital of a country. In order to analyze the internal corporate dynamics in the competitive environment, close monitoring of the leading actors of the ecosystem is a necessity in strategic management. In this study, the use of the SNA method in the context of technology and engineering management is demonstrated and an analysis of the principal actors in the field of automated vehicle (AV) technology is undertaken.

Introduction

SNA can be defined as the process of trying to understand social structures by using networks and graph theory. It is a method that defines the components hosted by a network as nodes and the connections connecting these components as edges (Nooy et al., 2011). Visualizations, usually created in the form of sociograms, provide visual representations of the nodes and edges, reflecting network-related properties. This makes it easier to evaluate the details of the network (Carrington et al., 2005; Wasserman & Faust, 1994; Wasserman & Galaskiewicz, 1994). In this respect, it can be said that SNA and visualizations provide important advantages in placing the patterns in large datasets. The several indicators obtained through SNA can explain the different dimensions of these patterns. These metrics and parameters are important in uncovering and explaining relation networks. It is necessary to start the analysis of the metrics by discussing centrality values. Centrality values are indicators that are frequently used to interpret relationships between nodes in relationship networks. The most frequently used of these values are the degree, degree centrality, betweenness centrality, closeness centrality, low aggregate constraints (LACs), and higher aggregate constrains (HAC).

Network Analysis and Network Metrics

Degree centrality refers to the number of relationships an item in a network has. For instance, in author analysis, the degree centrality of an author directly depends on the co-authorship status of that author. Since the rate of the degree centrality is calculated directly in connection with the number of publications, with this indicator, measurements can be made about the authors, institutions, and countries that stand out in the network in terms of frequency (Otte & Rousseau, 2002).

In network analysis, nodes with high *betweenness centrality* are defined in terms of the deductions that multiple groups or nodes make to reach each other (Otte & Rousseau, 2002). In other words, these nodes are nodes that allow connections between groups. Nodes with high betweenness centralities (authors, institutions, or countries) can be interpreted as connection points within a network. They can be interpreted as nodes linked to issues outside the network's general characteristics (Chen et al., 2010)

Closeness centrality refers to the distance of an analyzed node from other nodes in the entire network. High closeness means that a node can reach other nodes in the network in one or several steps (Sabidussi, 1966). In other words, assuming that there is information flow between two nodes, information is transmitted from this shortest path much earlier than other paths (Otte & Rousseau, 2002). In addition, modularity and silhouette values can be used in the evaluation of a network.

Modularity refers to the value indicating how sharp the clustering can be in the network structure created. Accordingly, modularity takes a value between 0 and 1, and low modularity values indicate that the network cannot be clustered very clearly, while high modularity values indicate a well-structured network (Brandes et al., 2007). In this respect, it is worth noting that the modularity value gives information about the adequacy of the network for analysis (Chen et al., 2010).

The *silhouette* value is used to determine the separation of clusters from each other (Campello & Hruschka, 2006). This metric shows the cleavage levels of the clusters in the created network. This indicator can take values between −1 and 1. A cluster with a silhouette value of 1 means that it is completely separated from other clusters (Chen et al., 2010).

Density refers to the number of edges, expressed as a ratio of the maximum possible number of edges in the network. It can take values from 0 to 1. This metric can be used in the context of egocentric or sociocentric analyses. In egocentric networks, it allows the calculation of the density of the connections around a node. In sociocentric networks, it measures the density of the graph and the restrictions on the members of the network (Scott, 2000).

Diameter refers to the longest path in the network. It indicates the width of the network, that is, the distance between the two most distant nodes (Kadry & Al-Taie, 2014).

The *clustering coefficient* is a measure of the degree of clustering tendencies in a network.

In co-authorship networks, some authors, institutions, or countries may be better positioned in a network to create co-authoring links. In this respect, these parameters can also be ranked in terms of their positioning using TM and network values. Nodes with low restrictions (LACs) have more flexible structures in terms of mobility, whereas for large nodes (HACs) with restrictions, it is the opposite.

Technology Management and Social Network Analysis

In a globalizing world, it can be said that countries are divided into two: technology-creating countries and technology-using countries. As the world rapidly shifts to technology-intensive sectors, the added value of the sectors that need human power is rapidly decreasing. Information technology, electronics, communication technology, biotechnology, superconductors, new materials, software, and robotic automation technology are the technologies of the future. Employment in the service sector in developed countries is increasing rapidly compared to the people working in industrial sectors. When technology is transferred on a sectoral basis, it is possible to benefit from all the opportunities of technology and engineering management through a strategy of accessing new markets with international partners. Producing technology and exporting know-how can be considered to be among the most significant challenges of our time. The competitiveness of the information age is the main idea of system design, which involves developing technology management strategies with which it is possible to find and support the creative people of a country and create an atmosphere of science with projects of economic value.

SNA, which has important advantages in its ability to reveal the patterns in big datasets, is also an important tool that can be used to fill the gap at this point. SNA is used in technology and engineering management to identify themes, trends, and relationships, like in bibliometric studies (Pilkington & Teichert, 2006; Weng, 2014). In addition to this, SNA is used to map the intellectual structures of modern technology management (Duan, 2011; Still et al., 2014) and to reveal the basic features of a specific technology domain (Lee & Su, 2011). Efforts to develop methods for the analysis of innovation networks using this method are frequently discussed in patent analysis (Han et al., 2012; Jun & Park, 2013). SNA is also used for the identification of the emerging areas in information and communication technologies and for keyword analysis (Khan & Wood, 2015). Building on the previous research for these purposes, evaluations were made using network metrics for three principal actors (authors, institutions, and countries) conducting research in automated vehicle (AV) technologies, as well as analyses of trends and collaboration patterns. The author network, institution network, and country network, which constitute the three main networks subjected to analysis, and the general characteristics of these networks are presented in Table 3.1. A high number of connections between nodes in a network indicates the density of that network. The entities (nodes) defined in a network can have a maximum number of connections corresponding to the number of all the elements in the network.

In our research, besides the basic centrality metrics, other metrics that are frequently used in determining the roles taken by network elements were also used. If the analyzed network is not a directional network, the eigenvector centrality value has the same degree as the hub value. In light of this, only hub values were used in our study (Nooy et al., 2018).

Table 3.1 *Summary of automated driving networks: network-level properties*

Networks	Nodes/ edges	No. of components	Density	Average degree	Diameter	Clustering coefficient	No. of edges in the minimum spanning tree
Author network	428/797	68	0.010	3.724	19	0.638653398	360
Institution network	589/1 357	50	0.008	4.608	13	0.1781940460	539
Country network	108/485	3	0.061	6.556	6	0.387554765	105

Author Network

In this section, in which we measure the performance of researchers in automated vehicles technologies based on network analysis, the rankings based on network metrics are given in tables. In this way, it is easier to keep track of the researchers' rankings according to metrics (Table 3.2).

The authors from Table 3.2 ranked as the top 20 authors in terms of centralities, hubs, and structural holes. In terms of degrees, Nils Appenrodt had the highest number of connections with other authors in the network. Despite not ranking at the top for betweenness, this author's high hub and LAC placements both demonstrate their strong connections to other important AV scholars. With these indicators, it is possible to say that an author is a network element open to collaboration within their ability to use their position in the network. Other important authors and network values are presented in Table 3.2 and visualized in Figure 3.1.

As a nascent research domain, the structure of the author network will tend to evolve. In other words, multiple but small collaboration maps can be observed in the author collaboration network. In this respect, it can be observed that researchers such as Dongpu Cao, Christoph Stiller, and Nils Appenrodt have taken on prominent profiles. It is possible to say that the leadership characteristics of these researchers come to the fore in AV research through their network values.

Institution Network

When the SNA that was used for the authors was repeated for institutions, it was observed that Carnegie Mellon University was the institution with the most

Table 3.2 Top 20 authors in terms of betweenness centralities, hubs, degree, and structural holes

Degree	Betweenness	Hubs	LAC	HAC
Appenrodt, Nils	Cao, Dongpu	Appenrodt, Nils	Cao, Dongpu	Rossetti, Rosaldo J. F.
Stiller, Christoph	Litkouhi, Bakhtiar	Hahn, Markus	Appenrodt, Nils	Naito, Takashi
Brenk, Carsten	Khajepour, Amir	Brenk, Carsten	Dietmayer, Klaus	Wu, Nan
Cao, Dongpu	Chen, Long	Klappstein, Jens	Stiller, Christoph	Revilloud, Marc
Franke, Uwe	Lauer, Martin	Bloecher, Hans-Ludwig	Brenk, Carsten	Wang, Bing
Knoeppel, Carsten	Dolan, John M	Sailer, Alfons	Lauer, Martin	Cazorla, Francisco J.
Enzweiler, Markus	Zhao, Huijing	Dickmann Juergen	Knoll, Alois	Breuel, Gabi
Herrtwich, Ralf G.	Davoine, Franck	Knoeppel, Carsten	Kurt, Arda	Xiao, Zhongyang
Hahn, Markus	Ibanez-Guzman, Javier	Enzweiler, Markus	Dickmann, Juergen	Lee, Jong Min
Stein, Fridtjof	Gong, Jianwei	Herrtwich, Ralf G.	Franke, Uwe	Oh, Kwang Seok
Lauer, Martin	Laugier, Christian	Stein, Fridtjof	Chen, Long	Lima Pedro F.
Pfeiffer, David	Triebel, Rudolph	Pfeiffer, David	Litkouhi, Bakhtiar	Himmelsbach, Michael
Strauss, Tobias	Appenrodt, Nils	Franke, Uwe	Strauss, Tobias	Burger, Patrick
Litkouhi, Bakhtiar	Ozguner, Umit	Stiller, Christoph	Schreiber, Markus	Lee, Minchul
Schneider, Nicolai	Siegwart, Roland	Muntzinger, Marc	Dolan, John M.	Tateno, Shigeyuki
Takeuchi, Eijiro	Sun, Zhenping	Schreiber, Markus	Xu, Xin	He, Xiangkun
Takeda, Kazuya	Chen, Hong	Strauss, Tobias	Knoeppel, Carsten	Sawabe, Taishi
Klappstein, Jens	Gilitschenski, Igor	Dickmann, Juergen	Enzweiler, Markus	Hagita, Norihiro
Knoll, Alois	Takeda, Kazuya	Lauer, Martin	Herrtwich, Ralf G.	Kanbara, Masayuki
Dietmayer, Klaus	Knoll, Alois	Schneider, Nicolai	Stein, Fridtjof	Deguchi, Daisuke

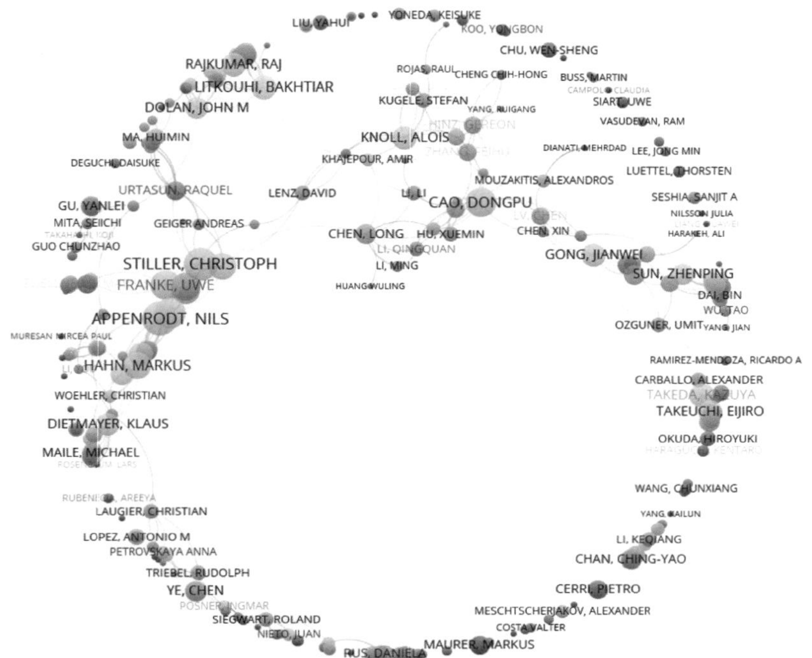

Figure 3.1 *Author network (degree)*

connections in terms of network values, showing the dominating influence of the Carnegie Mellon University Navigation Laboratory (Navlab), which was established at the university for studies on this subject. It also is an institution that enables the establishment of relations with other institutions in the network in terms of betweenness values. In other words, Carnegie Mellon University stands out as a critical institution for the AV network. Author network analysis showed that the network is broken up by several isolated author sets working in silos. It is possible to say that the diameter of the co-authoring network is small, and the cluster coefficient is high. This shows that the authors who contribute to the field of AVs have a high tendency to form groups. It is worth noting that, unlike the author network, the diameter of the institution network with 589 institutes is lower than that of the author network, while the cluster coefficient is low. Similarly, the institutions included in the structural analysis's list of well-positioned institutions that benefit from the network can be interpreted in terms of future developments in the field of AVs. There is strong evidence for network metrics in reviews of institutions' research performance.

It is possible to say that research institutions in the US, Europe, China, and Korea are in positions to establish intensive collaboration (Figure 3.2).

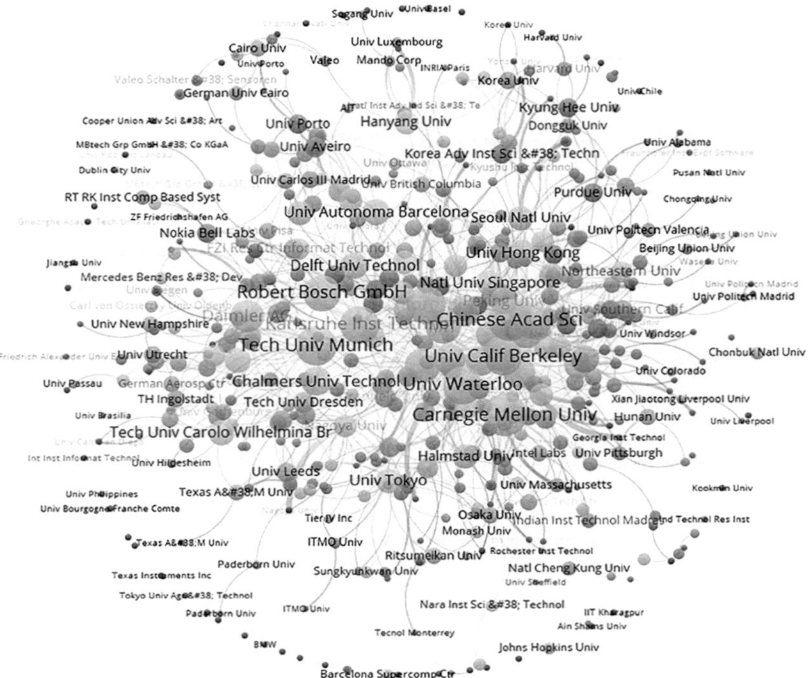

Figure 3.2 *Institution network (degree)*

It can be seen that the University of Waterloo plays an important role in mediation for information dissemination in terms of network values. In other words, the University of Waterloo can be considered a perfect center (Table 3.3). Considered in terms of LAC values, it can be seen that institutions that are open to cooperation can be easily determined. Institutes with high HAC values, such as Carnegie Mellon University, have reached a sufficient saturation level with the connection points they have already established in the AV network.

When we examine the productivity and importance of countries in the network, it can be seen that the US, Germany, and China provide the necessary infrastructure both in terms of the number of relationships in the network (degree) and the other nodes in the network to be connected (Table 3.4). We can say that intra-institutional collaborations are made with other sub units within the same institution. In fact, it would be useful to note that an intro-

Table 3.3 Top 20 institutions in terms of betweenness centralities, hubs, degree, and structural holes

Degree	Betweenness	Hubs	LAC	HAC
Carnegie Mellon University	University of Waterloo	Nagoya University	University of Waterloo	University of Western Australia
Chinese Academy of Sciences	Massachusetts Institute of Technology	University of Tokyo	Swiss Federal Institute of Technology	RT RK Institute for Computer-based Systems
Technical University of Munich	Carnegie Mellon University	Ohio State University	Stanford University	University of Novi Sad
Karlsruhe Institute of Technology	Technical University of Munich	Southwest Jiaotong University	Technical University of Munich	Nagoya University
University of Waterloo	Karlsruhe Institute of Technology	Toyota Technical Development Corporation	Carnegie Mellon University	Toyota Technical Development Corporation
Stanford University	Stanford University	Konkuk University	University of California, Berkeley	Doshisha University
University of California, Berkeley	Chinese Academy of Sciences	Doshisha University	Robert Bosch GmbH	Ohio State University
Robert Bosch GmbH	Tsinghua University	Denso	Chinese University of Hong Kong	Georgia Institute of Technology
Tsinghua University	Robert Bosch GmbH	Tier IV	Karlsruhe Institute of Technology	Hanyang University
Massachusetts Institute of Technology	Swiss Federal Institute of Technology	Ono Sokki	University of Maryland	Wuhan University
Wuhan University	University of California, Berkeley	National Traffic Safety and Environment Laboratory	Peking University	Central South University
Swiss Federal Institute of Technology	Hanyang University	Ritsumeikan University	Zhejiang University	University of Waterloo
Zhejiang University	Chalmers University of Technology	Chinese Academy of Sciences	National University of Defense Technology	Carnegie Mellon University

Degree	Betweenness	Hubs	LAC	HAC
Daimler AG	Wuhan University	Osaka University	Shanghai Jiao Tong University	Massachusetts Institute of Technology
University of Michigan	Zhejiang University	University of Waterloo	Tsinghua University	Polytechnic University of Turin
Sun Yat-sen University	Delft University of Technology	Hong Kong Polytechnic University	Daimler AG	Tsinghua University
University of Maryland	University of Tokyo	Sun Yat-sen University	University of Electronic Science and Technology of China	Hong Kong Polytechnic University
National University of Defense Technology	Seoul National University	Georgia Institute of Technology	University of Hong Kong	General Motors
Shenzhen University	University of Michigan	University of Science and Technology of China	Chalmers University of Technology	Seoul National University
Chalmers University of Technology	University of Maryland	Toyota Central R&D Labs	University of Texas at Austin	Zhejiang University

verted R&D activity pattern was also seen in the emerging technologies cases examined in the previous sections of the study. To reiterate, this situation, as in other emerging technologies, causes AV technologies to be subject to the conditions of early research, making it necessary to rely on limited areas of expertise and to choose the way to research them with internal resources due to the high potential of the technology. From a country-level network analysis perspective, the US, Germany, China, France, Italy, Spain, and the UK are key players, with the US playing the primary center role in the network; US institutions collaborate heavily with other institutions and universities in the US and with those in Germany, China, and Spain.

Discussion

In their automatic control systems, AVs, which can be defined as vehicles that can navigate without a driver's intervention by independently perceiving the road, traffic flow, and environment, make use of technologies and techniques such as radar, lidar, GPS, odometry, and computer vision. The first example of autonomous vehicles, the historical development of which dates to the

Table 3.4 *Top 20 authors in terms of betweenness centralities, hubs, degree, and structural holes*

Degree	Betweenness	Hubs	LAC	HAC
United States	United States	People's Republic of China	Iran	People's Republic of China
Germany	Germany	United States	Philippines	United States
People's Republic of China	People's Republic of China	Germany	Latvia	Germany
France	Spain	Canada	Vietnam	Canada
Italy	France	United Kingdom	United Arab Emirates	United Kingdom
Spain	Austria	Japan	Zimbabwe	Japan
United Kingdom	Canada	South Korea	Argentina	South Korea
Australia	United Kingdom	Australia	Malta	Australia
Canada	Italy	Singapore	Romania	Singapore
South Korea	South Korea	Spain	Kuwait	Spain
Japan	Brazil	Netherlands	Poland	Netherlands
Netherlands	Sweden	Italy	Croatia	Italy
Sweden	Egypt	France	Greece	France
Austria	Greece	Sweden	Mexico	Sweden
Belgium	Japan	Switzerland	Finland	Switzerland
Singapore	Serbia	Qatar	Cameroon	Qatar
Switzerland	Netherlands	Israel	Venezuela	Israel
India	Australia	Belgium	Brazil	Belgium
Ireland	Poland	Hungary	Malaysia	Hungary
Portugal	Portugal	India	Morocco	India

1920s, came to light with the Navlab and Autonomous Land Vehicle (ALV) projects carried out by Carnegie Mellon University in 1984. In 1987, the Eureka Prometheus project, which was carried out jointly by Mercedes-Benz and Bundeswehr University, turned into a Pan-European project rather than an individual project and set an example for R&D studies based on a cooperation model involving many universities and automobile manufacturers.

Scientific studies of AV technologies, which clearly have very high potential, were determined based on SNA in order to identify the leading actors in terms of authors, institutions, and countries. The roles of each actor in the network analysis were compared in terms of their centrality values. Thanks to the data obtained, authors, institutions, and countries that are open to relationships in terms of cooperation practices were identified. It is hoped that the analyses exemplified in this research can be used to guide decision makers in technology and engineering management.

ELECTRIC VEHICLES: DOMINANT ACTORS AND MAIN RESEARCH DOMAINS

In this section, the focus is on the SNA of the research on electric vehicles. In the analysis, comparisons were made for the centrality metrics and the leading actors were determined. At the same time, calculations were undertaken regarding the theory of structural holes (aggregate constraints) for the identification of actors who are open to development, and inferences were made on the use of the results obtained in the TM. It was observed that the electric vehicles, lithium-ion batteries, and hybrid electric vehicles clusters were the top five clusters in the subject areas revealed in the cluster analysis. Tsinghua University shows high levels of performance in terms of degree, betweenness, hubs and HAC values. As an author, Ouyango Minaggao (Tsinghua University) plays an important role in research about electric vehicles in terms of the level of connectivity. The US performs significantly better in terms of the degree of connectivity (betweenness) and, among other actors in the network, China can be seen as an important actor in terms of HACs.

Introduction

Vehicles producing energy through fossil fuels are still the most frequently used type in today's market. These vehicles operate using traditional energy technologies and, as a result of their activities, they cause environmental damage in various respects, especially through carbon monoxide emissions. Various studies on renewable energy options, such as solar and wind energy, have been carried out to avoid the damage caused by producing energy through fossil fuels. This trend has started to be observable in automobile technologies as well. Vehicles were previously produced with hybrid technology, manufactured using both fossil fuels and in electricity-powered ways (Freedonia Group, 2006, 2014; International Energy Agency, 2007a, 2007b). Over the past decade or so, we have begun to see the production of fully electric cars. Although many issues related to electric cars have been discussed in the literature and in industry, the biggest challenge concerns battery technologies (Rand et al., 1998).

A number of studies on renewable energy sources have shown an increasing trend in issues related to global warming, air pollution, and the effect on human health. AV technology is a subject frequently researched in the transportation domain (Hüls & Remke, 2017; Ramirez Barreto et al., 2018; Tian et al., 2018; Zhang et al., 2015). Research on electric cars has focused on both the technology itself and on the use of SNA to closely examine the evolution of these studies (Sun et al., 2018).

In this part of the study, using the bibliographic data for publications on electric vehicles in the international scientific literature, determinations based on SNA were made. As described in the SNA section of the study, the dominant actors in the technology domain of electric vehicles were determined based on their centrality measures. In addition, emerging research clusters were identified, and the network of cluster elements was visualized.

Results

In determining the clusters, common concepts in the titles, abstracts, and keywords of scientific documents were used. An algorithmic technique called latent semantic analysis (LSA) was used. LSA is used in natural language processing, particularly in distribution semantics, to show the relationships between a set of documents and the terms they contain. It is used to reveal the common characteristics of a document or documents by producing a series of concepts related to the documents and terms (Figure 3.3).

According to the cluster analysis results, the network can be divided into 17 co-citation clusters. Interpretation and verification of the consistency within the clusters obtained has been discussed. In this way, how well each object in the cluster was classified and how similar the objects belonging to the cluster were compared to their cluster were determined (Petrovic, 2006). These clusters are labeled by index terms from their own criteria. The largest six clusters are summarized in Table 3.5.

The largest cluster (cluster 0) had 301 members and a silhouette value of 0.818. It was labeled as ‘smart grid’ by LLR and ‘electric vehicles’ by term frequency–inverse document frequency (TFIDF). The second largest cluster (cluster 1) had 234 members and a silhouette value of 0.894. It was labeled as ‘electric vehicle’ by LLR and ‘lithium-ion batteries’ by TFIDF. The third largest cluster (cluster 2) had 174 members and a silhouette value of 0.855. It was labeled as ‘lithium-ion batteries’ by LLR and ‘state’ by TFIDF. The fourth largest cluster (cluster 3) had 166 members and a silhouette value of 0.839. It was labeled as ‘hybrid electric vehicle’ by LLR and ‘hybrid electric vehicles’ by TFIDF. The fifth largest cluster (cluster 4) had 149 members and a silhouette value of 0.796. It was labeled as ‘life-cycle assessment’ by LLR and ‘electric vehicles’ by TFIDF. The sixth largest cluster (cluster 5) had 148 members and a silhouette value of 0.808. It was labeled as ‘hybrid energy storage system’ by LLR and, ‘electric vehicles’ by TFIDF.

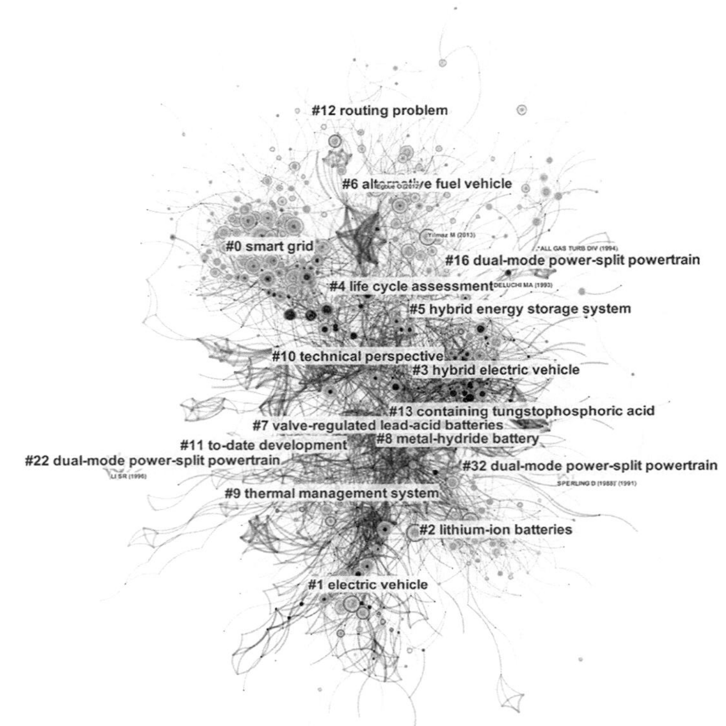

Figure 3.3 Clusters (mutual information)

Table 3.5 Cluster details: electric vehicles

Cluster ID	Size	Silhouette	Label (term frequency–inverse document frequency)	Label (LLR)	Mean cited year
0	301	0.818	Electric vehicles	Smart grid	2012
1	234	0.894	Lithium-ion batteries	Electric vehicle	2008
2	174	0.855	State	Lithium-ion batteries	2011
3	166	0.839	Hybrid electric vehicles	Hybrid electric vehicle	2008
4	149	0.796	Electric vehicles	Life-cycle assessment	2011
5	148	0.808	Electric vehicles	Hybrid energy storage system	2010

Author activity

The workflow used in the previous section was used to identify the authors. The author network is shown in Figure 3.4.

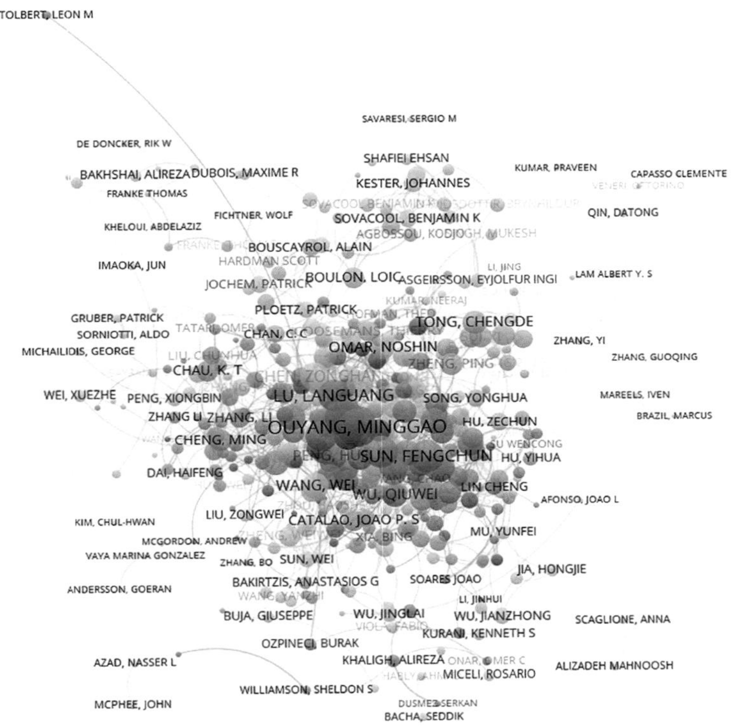

Figure 3.4 Author network (degree)

The list of the top 20 authors ranked in terms of network metrics is given in Table 3.6. It can be observed that Minggao Ouyang (Tsinghua University) plays an important role in research on electric vehicles in terms of the level of connectivity in this network, the relationships with other nodes, and their mediation values. Likewise, Jianqiu Li from the same university demonstrates significant performance in terms of connectedness and mediation values. Although both authors are unlikely to withdraw from the network, when the HAC values are examined, it can be observed that these authors exhibit a situation suitable for new collaboration ties. In other words, the higher the total constraint, the less 'freedom' is required for a person to withdraw from existing bonds or take advantage of structural holes. It has been shown that individuals with low constraints have more successful careers in an organization. In this regard, the LAC value can be used to identify nodes that do not have sufficiently high network values in terms of connectivity but that need to be considered in terms of success (Table 3.6).

Table 3.6 Top 20 authors in terms of centralities, hubs, and structural holes

Degree	Betweenness	Hubs	HAC	LAC
Ouyang, Minggao	Ouyang, Minggao	Ouyang, Minggao	Longo, Michela	Hu, Xiaosong
Li, Jianqiu	Hu, Xiaosong	Li, Jianqiu	Marinelli, Mattia	Wang, Hao
Sun, Fengchun	Jiang, Jiuchun	Lu, Languang	Jain, Praveen	Peng, Huei
Hu, Xiaosong	Lin, Zhenhong	Han, Xuebing	Dai, Haifeng	Yang, Xiao-Qing
Lu, Languang	Mi, Chunting Chris	Xu, Liangfei	Antunes, Carlos Henggeler	Amine, Khalil
Han, Xuebing	Zhu, Jianguo	Feng, Xuning	Golkar, Masoud Aliakbar	Chen, Zheng
Xiong, Rui	Amine, Khalil	Xu, Liangfei	Xu, Kun	Yang, Chao
Amine, Khalil	Ploetz, Patrick	Zheng, Yuejiu	Xu, Guoqing	Zhu, Jianguo
Jiang, Jiuchun	Wang, Wei	Song, Ziyou	Imura, Takehiro	Wang, Michael
Chen, Zonghai	Axsen, Jonn	Hofmann, Heath	Xu, Zhiwei	Pecht, Michael
He, Hongwen	Wang, Peng	Fang, Chuan	Devetsikiotis, Michael	Wang, Bin
Sun, Yang-Kook	Zhang, Lei	He, Xiangming	Bayram, I. Safak	Lin, Zhenhong
Feng, Xuning	Chen, Long	Lehnert, Werner	Sun, Zechang	Xiong, Rui
Mi, Chunting Chris	Wang, Hao	Zheng, Yuejiu	Paffumi, Elena	Li, Li
Van Mierlo, Joeri	Wu, Qiuwei	Wang, Hewu	Pahlevaninezhad, Majid	Liu, Xiangdong
Wang, Wei	Zhu, Chunbo	Hou, Jun	Pantic, Zeljko	Sun, Fengchun
He, Xiangming	Li, Kang	Ouyang, Minggao	Bradley, Thomas H	Sun, Yang-Kook
Zhang, Lei	Li, Jianqiu	Zhang, Xiaowu	Fujimoto, Hiroshi	Keoleian, Gregory A.
Wang, Junmin	Wang, Junmin	Wang, Li	Onat, Nuri Cihat	Wu, Qiuwei
Shen, Weixiang	Murgovski, Nikolce	Wang, Junmin	Wu, Jinglai	Zhang, Jianhua

Institution activity

Another analysis revealed the productivity of the institutions in the studies on electric vehicles and the affiliation information of the authors (Figure 3.5).

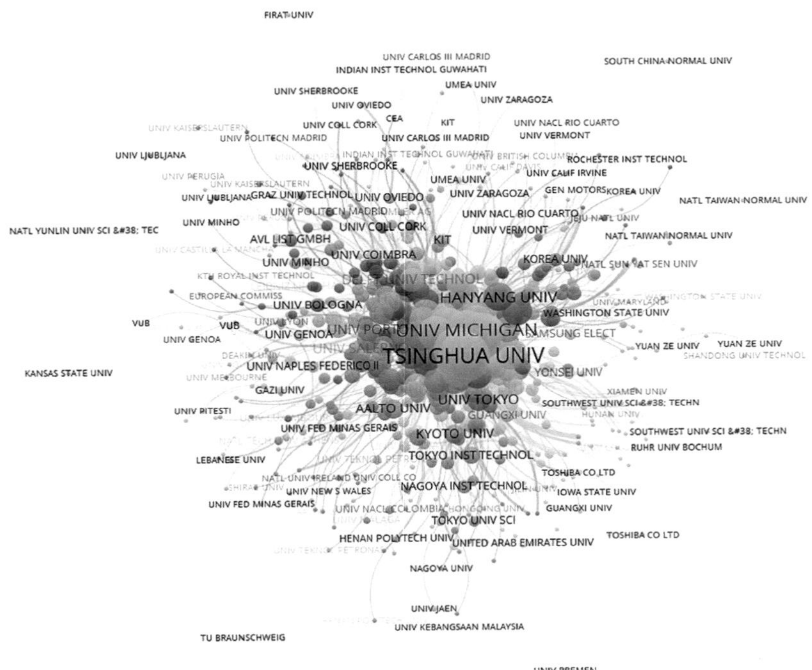

Figure 3.5 *Institution network (degree)*

According to the table created for the centrality metrics, Tsinghua University shows a high level of performance in terms of degree, betweenness, hubs and HAC values. It is worth noting that the three representatives of European universities (Aalborg University, Chalmers University of Technology, and the Technical University of Denmark) are the institutions to follow in terms of LAC values (Table 3.7). The results can be expected to be parallel to those in the institutional effectiveness analysis since the data for the actors in the author effectiveness table were derived from the affiliation information. However, the table can be used to closely examine the institutions to which the authors belong and to determine the role of institutions in issues such as technological knowledge redundancy and status.

Table 3.7 *Top 20 institutions in terms of degree centrality, hubs, and structural holes*

Degree	Betweenness	Hubs	HAC	LAC
Tsinghua University	Tsinghua University	Tsinghua University	Tsinghua University	Aalborg University
Chinese Academy of Sciences	Chinese Academy of Sciences	University of Michigan	University of Michigan	Chalmers University of Technology
University of Michigan	University of Michigan	Beijing Institute of Technology	Beijing Institute of Technology	Technical University of Denmark
Argonne National Laboratory	Argonne National Laboratory	Chinese Academy of Sciences	University of California, Berkeley	Huazhong University of Science and Technology
Beijing Institute of Technology	University of California, Berkeley	Argonne National Laboratory	Chongqing University	Massachusetts Institute of Technology
University of California, Berkeley	Massachusetts Institute of Technology	University of California, Berkeley	Argonne National Laboratory	University of Texas at Dallas
Zhejiang University	University of Waterloo	University of the Chinese Academy of Sciences	Xi'an Jiaotong University	University of Wollongong
Shanghai Jiao Tong University	Aalborg University	Collaborative Innovation Center of Electric Vehicles in Beijing	Stanford University	North Carolina State University
University of Waterloo	Zhejiang University	Chongqing University	University of Waterloo	Georgia Institute of Technology
Massachusetts Institute of Technology	Seoul National University	Ford Motor Company	Jilin University	Hong Kong Polytechnic University
Nanyang Technological University	Delft University of Technology	Beihang University	Zhejiang University	University of New South Wales
Georgia Institute of Technology	Technical University of Denmark	Xi'an Jiaotong University	Beihang University	City University of Hong Kong

Degree	Betweenness	Hubs	HAC	LAC
Technical University of Denmark	Beijing Institute of Technology	Stanford University	Shanghai Jiao Tong University	Southeast University
Chongqing University	Chalmers University of Technology	Tianjin University	RWTH Aachen University	Hunan University
Huazhong University of Science and Technology	Nanyang Technological University	Oak Ridge National Laboratory	Hanyang University	Argonne National Laboratory
Oak Ridge National Laboratory	Shanghai Jiao Tong University	Forschungszentrum Jülich	Beijing Jiaotong University	Polytechnic University of Milan
Stanford University	Katholieke Universiteit Leuven	Massachusetts Institute of Technology	Nanyang Technological University	University of Waterloo
Hanyang University	Georgia Institute of Technology	Shanghai Jiao Tong University	Massachusetts Institute of Technology	Karlsruhe Institute of Technology
Seoul National University	Polytechnic University of Milan	University of Waterloo	Ohio State University	Arizona State University
Beijing Jiaotong University	Hanyang University	Beijing Jiaotong University	University of Texas at Austin	Shanghai Jiao Tong University

In the network of research on electric vehicles, it can be observed that the important roles taken in terms of network metrics are distributed between the US, the UK, and China (Figure 3.6).

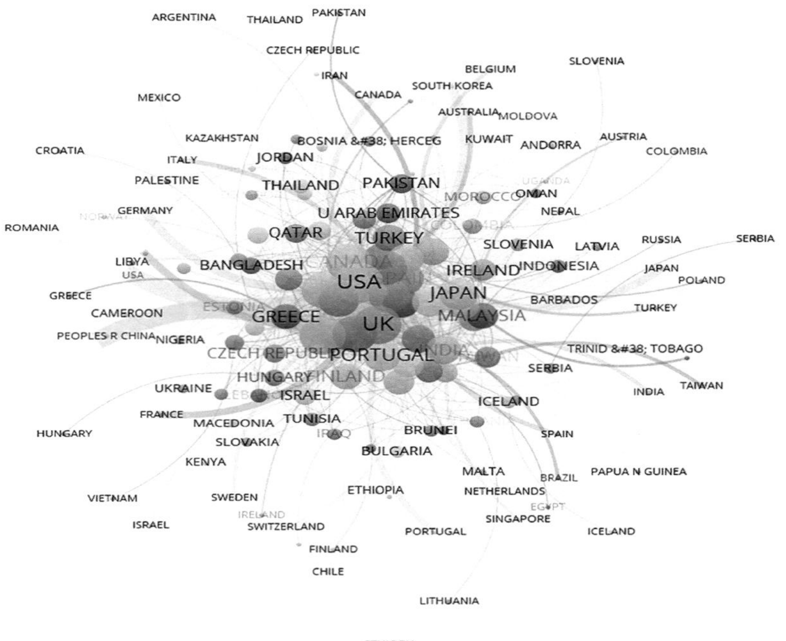

Figure 3.6 Country network

When the country network is examined closely, it can be seen that, unlike the result that emerges in the corporate network, institutions with US addresses stand out in terms of centrality values in the network. Although it can be seen that institutions with Taiwanese addresses are at the forefront in terms of institution performance, Taiwan is further away from the center of the network compared to countries such as the US, UK, and Canada. When we evaluate it from this point of view, it is clear that dealing with network visuals with detailed indicators as much as possible can provide micro-level information. In this context, while the US performs well in terms of the degree of connectivity (betweenness) with other actors in the network, China should be seen as an important actor in the network in terms of HAC values. Finland, Egypt, the Czech Republic, New Zealand, Greece, Switzerland, Denmark, Russia, and Chile can also be seen to be countries worth following (Table 3.8).

Table 3.8 Top 20 countries in terms of degree centralities and structural holes

Degree	Betweenness centrality	Hubs	HAC	LAC
United States	United States	People's Republic of China	People's Republic of China	Finland
United Kingdom	United Kingdom	United States	United States	Egypt
People's Republic of China	People's Republic of China	United Kingdom	United Kingdom	Czech Republic
Canada	Germany	Australia	Germany	New Zealand
Germany	France	Canada	South Korea	Greece
France	Canada	Germany	Canada	Switzerland
Australia	Australia	South Korea	Japan	Denmark
Italy	Spain	Singapore	Australia	Russia
Spain	Italy	Japan	France	Chile
Denmark	Denmark	France	Italy	Italy
Japan	South Korea	Italy	Cameroon	Norway
Netherlands	Switzerland	Denmark	Singapore	Israel
Portugal	Malaysia	Sweden	Portugal	Netherlands
Iran	Japan	Saudi Arabia	Ghana	Estonia
South Korea	Iran	Iran	Spain	Austria
India	Belgium	India	Iran	Malaysia
Sweden	Turkey	Netherlands	Kenya	Thailand
Brazil	Brazil	Norway	India	Sweden
Belgium	Egypt	Portugal	Sweden	Croatia
Turkey	Finland	Spain	Netherlands	Cyprus

In the table, in which the important members of the network can be clearly identified in terms of connectivity indicators, countries that have important roles in terms of intermediation values (HUB values) are also identified. Accordingly, it can be observed that countries such as the People's Republic of China, the US, and the UK play very important roles in electric vehicle research in terms of connectivity, centrality, and intermediation values, and the fact that these countries share the top positions of the HAC values list strengthens the findings on this issue.

SMART HOMES: CONNECTED DAILY LIFE OF HUMANKIND

Smart home technologies are growing in popularity and increasingly integrated into our lives day by day. Smart home systems will also affect daily life by providing various services that are convenient for human life and by enabling our homes to be equipped in terms of security and remote-control issues. Smart houses appear at first glance as a topic for which automation is intense. Efforts aiming to achieve maximum efficiency by saving energy intend to both leave more energy resources available to future generations and optimize the cost of using energy resources.

The desire to be connected has driven people into the habit of keeping in touch with technological tools and equipment such as smart technology. Human beings, who want to take advantage of technology, use the tools they have in every part of their daily lives, including the activities involved in daily home life. Developments such as sensor networks, connected devices, the Internet, and the IoT have also facilitated the usage functions of the tools used in daily life at home. In this section, the concept of the smart home is closely examined and the research on this subject is analyzed according to the principles of SNA. Major actors in the studies, such as researchers, institutions, and countries, were determined and the importance of these actors in the field was calculated using graph theory.

There are many studies in the literature showing that technology seems to be more of a tool than a purpose, since people think of technology as part of everyday life (Weiser & Brown, 1996, 1997). This approach, which is called calm technology, includes the principle that technology should only do its part after people adjust the technology to benefit them. In this way, it is possible for technology to come into play at every stage that people need it in daily life. From this point of view, it can be observed that many such technologies have been produced in many sectors and fields, such as wearable health technologies (Burmaoglu et al., 2018) and air conditioning systems that can be controlled over the Internet (Saputra & Lukito, 2017). Smart homes can be defined as the adoption of control systems, which are used in many areas of industry, to everyday home life (Alam et al., 2012). With home automation, these technologies are applied for the special needs and wishes of individuals. In other words, smart home technologies can satisfy the needs of residents, making their lives easier and offering a safer, more comfortable, and more economical life (De Silva et al., 2012).

Smart houses are structured so that each user can control the automatic home functions and systems remotely. One of the most important motivational factors in the transition to smart home automation is the desire to increase efficiency and save energy (Reinisch et al., 2011). Increases in energy costs for a normal family are often caused by unnecessary energy consumption. Smart home technologies can help reduce unnecessary energy drains such as heating and lighting in unused areas of the house, heating and cooling systems that are left on too high or inconsistently regulated, and blinds and drapes left open during the warmest and coldest times of day, thereby creating energy loss. The developments in such technologies have directly affected the technologies used in smart homes. People can remotely control the contents of refrigerators in their homes, their air conditioners, or their heating systems (Balta-Ozkan et al., 2013). In our study, life-cycle development and movement in smart home technologies were analyzed within the framework of increasing technological studies using SNA. In this way, the determination of key technologies and sub-technologies and researchers, institutions, and countries that dominate the field were made. General network statistics that appeared in the network analysis for the dominant actors are given in Table 3.9.

Table 3.9 *Network statistics*

Networks	Nodes/ edges	No. of components	Density	Average degree	Diameter	Clustering coefficient	No. of edges in the minimum spanning tree
Author network	418/742	97	0.009	3.550	5	0.515876293182373	321
Institution network	569/1 059	74	0.007	3.722	13	0.25416284799575806	495
Country network	130/693	3	0.083	10.662	6	0.4389883875846863	127

When the data were analyzed, we found a considerable growth rate for smart technologies. This growth is consistent with the exponential growth curve ($R^2 = 0.8774$). In this respect, it is possible to say that, while smart home technologies significantly increased in the period from 2010 to 2014 in accordance with the developing technological opportunities, they have enormously increased from 2015. As can be seen in Figure 3.7, the years 2014–2015 were a turning point for smart home technology research. This movement in smart home technologies cannot be considered separately from research on sensor networks and the IoT.

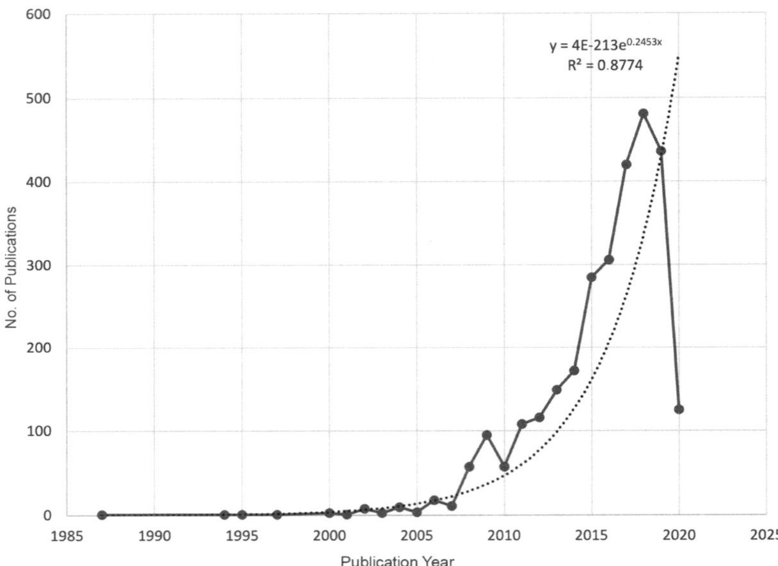

Figure 3.7 Literature growth for smart home publications

We examined the authors' collaborations and the network values for each author showing the level of connectivity in the network were calculated. The visualization made for the network values (degree) is given in Figure 3.8. In order to measure the scientific effectiveness of each author and their impact on the community, the total numbers of citations and publications along with the h-index values are also presented in Table 3.10.

Table 3.10 Author productivity

h-index	Author name	h-core citation sum	All citations	All articles
12	Cook, Diane J.	997	1 089	31
10	Schmitter-Edgecombe, Maureen	534	551	16
10	Nugent, Chris	445	482	24
10	Javaid, Nadeem	282	358	49
8	Bouzouane, Abdenour	125	188	35
8	Khan, Zahoor Ali	230	239	16
8	Bouchard, Bruno	125	181	34
7	Chen, Liming	1 004	1 005	9
7	Alrajeh, Nabil	220	226	8
6	Nugent, Chris D.	765	776	10

The table presents data that can be used both for identifying the dominant actors of the technology domain and for recruitment purposes. The same table can also be used as an important tool for the development of strategies in R&D planning or to shape the know-how transfer based on transfers between experts. In this way, it is possible to develop a data-based policy. The researchers whose individual performances were determined in the table were evaluated using the indicators based on SNA. For this, utilizing graph theory principles, each node was positioned in the network map according to the centrality values it represented. In the resulting map, important nodes in the network could be clearly shown (Figure 3.8).

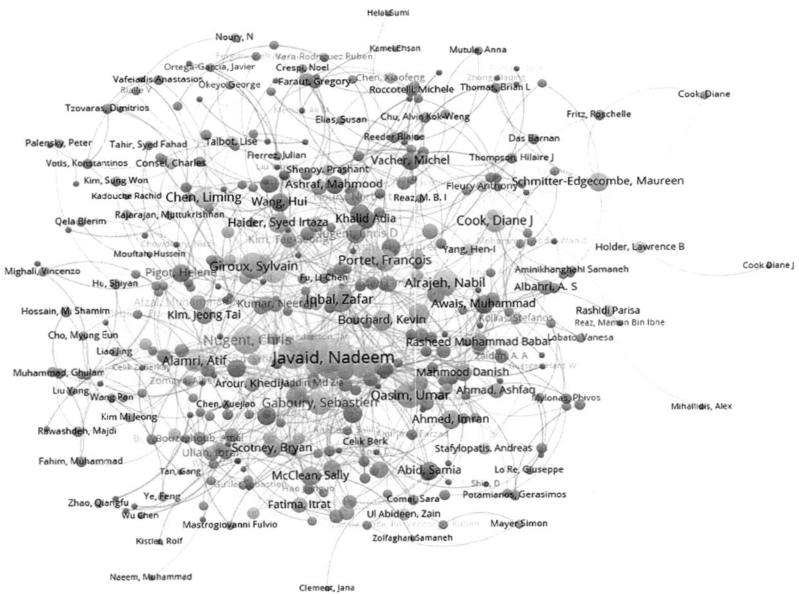

Figure 3.8 *Author network*

It is possible to see that Nadeem Javaid plays an important role in terms of both network values and productivity values. On the other hand, Abdenour Bouzouane, Bruno Bouchard, Diane J. Cook, and Zahoor Ali Khan play important roles in research on both kinds of smart productivities, and they are therefore the key researchers for smart home research. Researchers and their affiliated institutions are among the important indicators of investigation in terms of collaborations. In this context, when the performances related to the collaborations of the countries were examined, it could be observed that countries such as the US and China are prominent in terms of network values although the collaborations are limited in such countries (Figure 3.9).

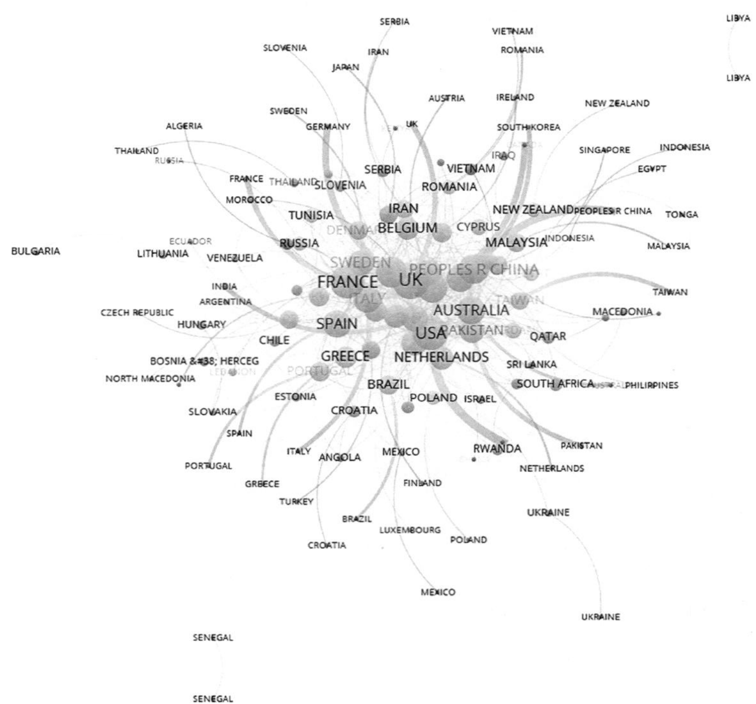

Figure 3.9 *Country network (degree)*

We have mentioned that countries' institutions tend to collaborate with each other. Examining the countries in terms of the pattern of collaboration, it is possible to summarize the situation as in Figure 3.9. As mentioned before, this situation can be attributed to reasons such as the technologies showing characteristics related to emerging technology fields with high potential for commercialization. A close examination of the networks allows the collaborations between Asian, American, and European Union countries to be monitored. As with the other emerging technologies examined in the previous sections of our study, it can be observed that cooperation patterns are limited to other units within the organization.

It is possible to observe that King Saud University is among the most important nodes of the network in terms of network values. On the other hand, Islamabad University and Alberta University are important in terms of their centrality values.

Overall, the last decade saw much of the progress in this technology. In addition to institutions from the US, China, Australia, and European Union countries, we can also see institutions emerging from countries like Saudi Arabia. It looks like these countries are trying to leapfrog with this technology. Smart cities, which countries have tried to put into effect recently, are emerging as a factor that causes smart infrastructure applications to come to the forefront in terms of network values in smart technology research clusters. Although they are expressed in the form of smart home technology at the micro-scale, it

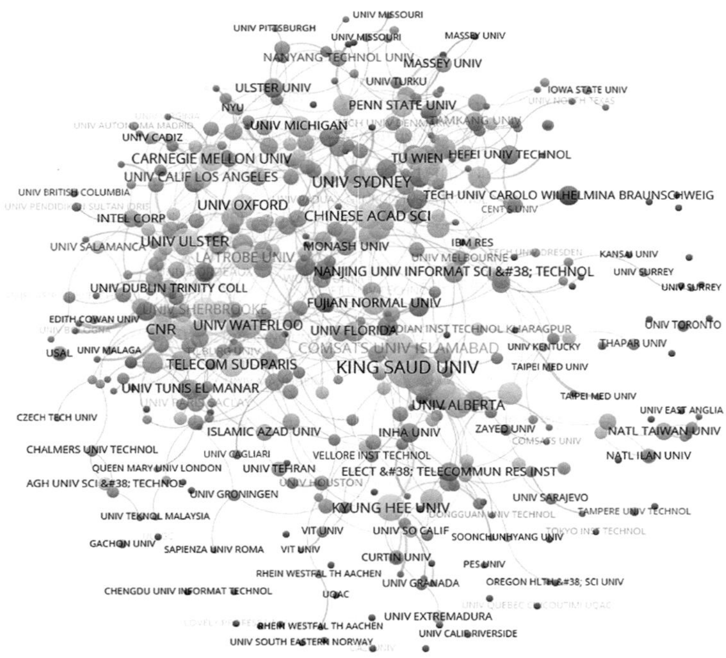

Figure 3.10 *Institution network (degree)*

is clear that there are some necessary conditions that smart homes must meet to realize the possibilities they have across all their aspects. Smart houses that need smart infrastructure facilities due to their environment and the ecosystem they create are defined as smart cities (Figure 3.10).

THE SOCIAL NETWORK ANALYSIS OF SPACE TRAVEL

Space travel and tourism is one of the biggest challenges of human history. This process, which was fueled by the competition between the US and the Soviet Union in the race to travel to the moon, previously included the efforts to launch and retrieve live satellites (Glass, 1974). Today, operations such as sending rockets to Mars and collecting samples are carried out (Köpping Athanasopoulos, 2019; Larina et al., 2019; Launius, 2008; Szocik et al., 2017). Human beings' engagement in off-planet research activities has often been fed by the competitive environment following the Second World War (Simmons et al., 2016). In this relentless race between the US and the Soviet Union, the primary goal of the countries involved was to send rockets to the moon. In this regard, it can be observed that, while previously only hardware objectives were targeted, today intensive work on human travel is being carried out (Blue et al., 2017; Popova et al., 2020; Sannita et al., 2006). It can be observed that some of these studies focus on energy harvesting in space (Le et al., 2015). The general situation and the last point reached in space travel research, where such wide-ranging and high-cost studies are carried out, provide critical data for the determination of the progress made, especially in the context of technology and engineering management. In this respect, in order to invest in the right technology and the right R&D activities, it was considered that it would be useful to examine the research and development activities carried out in the field of space travel with TM in depth. With this information, SNA was conducted to determine the technological progress in the field of space travel. The data obtained were processed according to the principles of bibliometrics and scientometrics, which were turned into a TM application thanks to a mixed method. Bibliometrics and scientometrics combined with SNA can be used for a variety of purposes (Daim & Yalçin, 2019; Pelicioni et al., 2018; Yalçin et al., 2020).

Findings

In our study, we examined the scientific studies on space travel according to the principles of SNA and graph theory. As a result of this examination, it was possible to determine the technological advances in the relevant subjects and to identify the sub-technologies for space travel that are being studied intensively. In this respect, we intended to identify the leading actors in the field of space travel according to their level of competence and to determine the prominent people, institutions, and countries in this framework. To obtain a dataset, an online search was performed in all databases of WoS for the years from 1900 to 2020 and the bibliographic data of the publications indexed in the 'space travel' area were compiled topically. Several tools, such as the R programming language, ggplot 2 library, hive visualization library, and MS Excel, were used in the analysis of the data. VOSviewer software was used for the visualization of social network maps. As a result of the analysis, the annual growth rate of space research literature was calculated to be 5.1 percent. The bibliographic data obtained were compiled from 604 different sources. While the number of single-author publications was 444, the number of authors per document was 0.359. The author collaboration index was calculated as 4.36. It can be seen that the rate of cooperation in the field of space travel is quite high. When we consider it in terms of contributing countries, the US leads the way in space travel. Japan takes the second place and Belgium, China, and Germany are also at the top of the list. The countries are given in Figure 3.11, which shows the most dominant institutions in terms of competence along with the keywords that are extensively studied. If the examined scientific publications are considered in terms of their priorities, the most important are biological risks arising from radiation in the space environment, individual health and performance during spaceflight/space travel, logistics resources for space life, and life support systems in space.

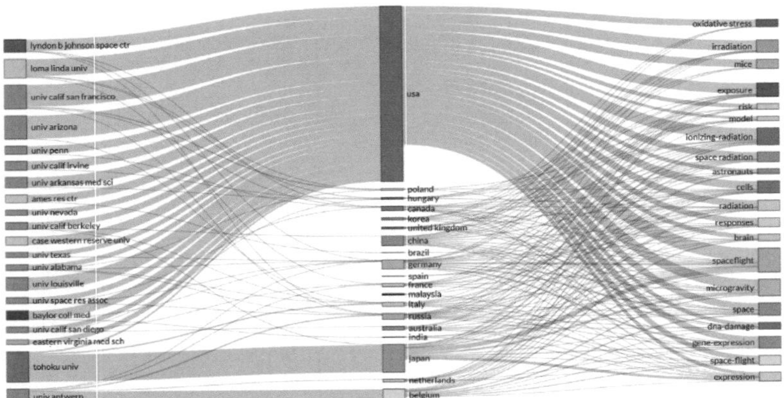

Figure 3.11 Three-field plot (institutions–countries–Keyword-Plus)

In a further analysis, looking at the clusters of scientific studies conducted in the field of space travel, the determination of the sub-technology domains in which the studies were mostly clustered was carried out. For this purpose, a method that facilitates the interpretation of categorical variables and graphically displays the similarities, differences, relationships, and changes of these variables in a less dimensional space was chosen. According to the visual obtained by this method, which is called multiple correspondence analysis (MCA), it is possible to examine space travel research under three main headings. Since human travel is now aimed at in space travel research, it is clustered around three main headings in the visual results of the MCA of the research on space travel (Figure 3.12). In this regard, it should be noted that space travel research is not solely composed of studies related to space science but still shows a research field dynamic that benefits from basic science research, especially biology and chemistry.

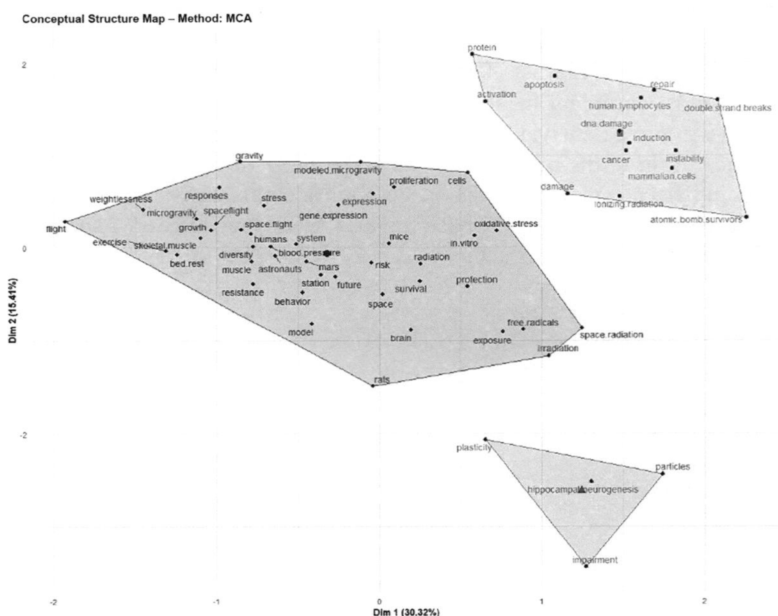

Figure 3.12 Keyword-Plus multiple correspondence analysis

In the co-word analysis, the subject areas studied intensely could be determined in more depth. Accordingly, it is worth mentioning that the terms ‘spaceflight’, ‘ionizing radiation’, and ‘exposure’ were the most important nodes in terms of network values. When these concepts are examined closely, it is possible to conclude that the research on space travel is still at an experimental level. It can be seen that these experimental studies focus more on radiation and its effects on living things (Table 3.11). The main purpose of space travel research, which still shows an early study profile, is human travel. For this purpose, studies that focus on the effects of space conditions on living things are frequently encountered. In fact, the obscurity in space is seen as the primary barrier to travel applications that must be eliminated. In this respect, it can be clearly seen that issues such as radiation and DNA structure are prioritized.

It can be observed that the determinations obtained in the MCA results are similar to the nodes revealed in the network analysis and the centrality values they possessed. As mentioned before, it can be clearly seen that focal points such as biological risks caused by radiation and health and performance during spaceflights come to the fore in terms of network values (Table 3.11). After determining the focal points of scientific research activities, the SNA method was used to determine the main actors in the relevant research ecosystems. In this context, a set of network values was calculated in order to determine the roles each network member assumes in the network (Table 3.11).

The finding that the US is at the center of space travel research should be seen as an acceptable result. In terms of network values and collaboration dynamics, it can be observed that institutions and organizations have structures that are prone to collaboration patterns, but these collaborations are mostly carried out with other departments in the same institution (Figure 3.13). The situation related to the budget allocated by countries and institutions for such research also shows itself in the results of SNA. In other words, it can be seen that countries with a high level of development stand out in terms of network values in the SNA of space travel research (Table 3.12).

In the analyses carried out in our study the centrality values and the correlation indicators in the network were prioritized. When the results of the SNA were evaluated, it could be observed that the US plays a very important role in space travel research, especially in terms of centrality values, while Germany, the UK, Italy, and Russia follow in terms of centrality values. In the visualization of the performances of the countries in Figure 3.14, when we look closely at the leading institutions and organizations in space travel research it is possible to summarize the situation as follows. The nodes located further away from the center of the network stand out as institutions where research that can be defined as secondary research rather than space research is conducted. In this respect, it should not be forgotten that the visual was created to describe the dominant actors of the field.

In terms of network values, NASA is the clear leader in space travel research. This is not surprising when we consider that it has undertaken the largest known research program.

Table 3.11 *Keyword-Plus co-occurrence analysis*

Label	All-degree partition	Hubs and authorities	Size of all domain	All degree	Weighted all degree	All closeness centrality	Betweenness centrality	Hub weights	Authority weights	Normalized size	Average distance	All proximity prestige	Aggregate constraint
Spaceflight	154	2	494	154	236	0.553	0.122	0.325	0.325	1	1.808	0.553	0.032
Ionizing radiation	134	2	494	134	215	0.529	0.067	0.271	0.271	1	1.891	0.529	0.031
Exposure	120	2	494	120	160	0.507	0.061	0.205	0.205	1	1.972	0.507	0.034
Irradiation	116	2	494	116	156	0.515	0.051	0.230	0.230	1	1.943	0.515	0.036
Microgravity	108	2	494	108	143	0.516	0.075	0.219	0.219	1	1.937	0.516	0.046
Space radiation	97	2	494	97	123	0.489	0.044	0.163	0.163	1	2.047	0.489	0.038
Mice	91	2	494	91	117	0.509	0.038	0.182	0.182	1	1.966	0.509	0.036
Space	91	0	494	91	105	0.521	0.071	0.153	0.153	1	1.921	0.521	0.035
Radiation	84	0	494	84	100	0.509	0.035	0.143	0.143	1	1.964	0.509	0.033
Cells	83	2	494	83	113	0.505	0.030	0.193	0.193	1	1.982	0.505	0.041
Responses	80	0	494	80	102	0.498	0.045	0.151	0.151	1	2.008	0.498	0.045
Spaceflight	79	2	494	79	107	0.503	0.032	0.157	0.157	1	1.990	0.503	0.043
Oxidative stress	79	2	494	79	102	0.499	0.030	0.171	0.171	1	2.002	0.499	0.041
Expression	78	0	494	78	96	0.496	0.040	0.144	0.144	1	2.014	0.496	0.039
Gene-expression	77	0	494	77	90	0.496	0.042	0.117	0.117	1	2.016	0.496	0.037
Brain	68	0	494	68	83	0.490	0.020	0.133	0.133	1	2.043	0.490	0.044
Astronauts	66	0	494	66	84	0.494	0.033	0.136	0.136	1	2.024	0.494	0.049
DNA damage	64	0	494	64	80	0.476	0.018	0.112	0.112	1	2.099	0.476	0.048

Label	All-degree partition	Hubs and authorities	Size of all domain	All degree	Weighted all degree	All closeness centrality	Betweenness centrality	Hub weights	Authority weights	Normalized size	Average distance	All proximity prestige	Aggregate constraint
Model	64	0	494	64	67	0.498	0.035	0.089	0.089	1	2.008	0.498	0.038
Activation	58	0	494	58	73	0.475	0.014	0.128	0.128	1	2.103	0.475	0.052

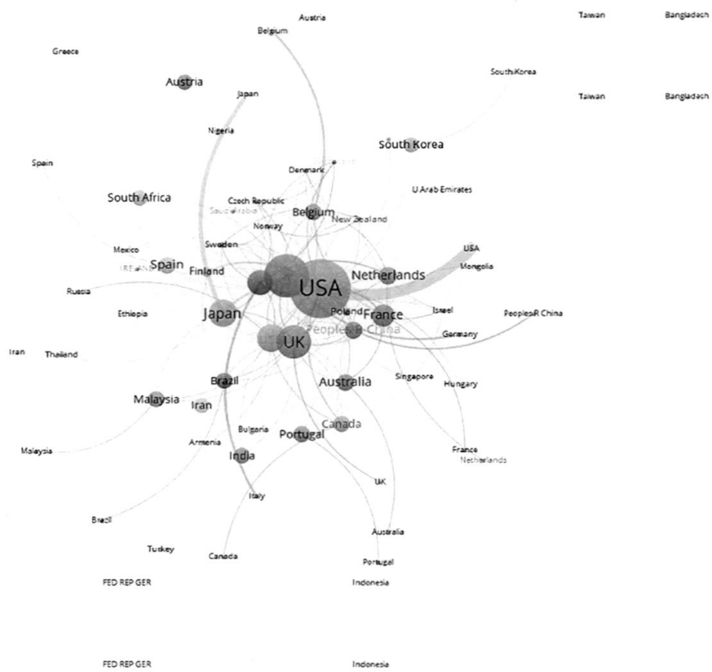

Figure 3.13 Country collaboration network

Discussion

Space research is one of the activities based on scientific research and development activities whose history gained momentum after the Second World War. The Soviet Union's launch of the Sputnik 1 artificial satellite into Earth's orbit on 4 October 1957, and then the US's landing on the moon with the Apollo 11 spaceship on 20 July 1969 are known as the beginnings of this space race. While this process, which was exacerbated by the race between Russia and the US in sending a shuttle to the moon, was considered as one that was only aimed at sing hardware to AY in the early period, much progress has been made in the process today. These advances have recently turned into attempts at manned space travel. In this respect, determining the main and sub-technology areas in which studies are concentrated, as well as the main actors and which countries, institutions, and persons prioritize such technologies has provided important information. The details provided by SNA and graph theory have shown that they can provide strategic information to decision makers within

Table 3.12 *Country collaboration network parameters*

Label	Input degree	Periodic strong components	All neighbors	All core partitions	Multi-level Louvain communities	Hubs and authorities	Size of all domains	All degree	All weighted degrees	All closeness centralities	Between-ness centrality	Hub weights	Authority weights	Normal-ized size	Average distance from all domains	All proximity prestige values	Aggregate constraint
United States	52	1	1	10	1	2	63	52	2056	0.566	0.330	0.705	0.709	0.887	1.571	0.565	0.588
Germany	38	1	2	10	3	2	63	38	295	0.523	0.181	0.035	0.035	0.887	1.698	0.522	0.223
United Kingdom	30	1	2	10	3	2	63	30	160	0.505	0.110	0.014	0.014	0.887	1.762	0.504	0.213
Italy	27	1	2	10	3	2	63	27	239	0.479	0.061	0.018	0.018	0.887	1.857	0.478	0.255
Russia	22	1	2	10	4	2	63	22	151	0.475	0.075	0.012	0.012	0.887	1.873	0.474	0.185
France	20	1	2	10	8	0	63	20	77	0.452	0.049	0.007	0.007	0.887	1.968	0.451	0.214
Japan	19	1	2	10	2	2	63	19	390	0.463	0.074	0.041	0.041	0.887	1.921	0.462	0.454
People's Republic of China	16	1	2	10	5	0	63	16	97	0.444	0.029	0.006	0.006	0.887	2.000	0.444	0.401
Netherlands	15	1	2	10	3	0	63	15	89	0.434	0.032	0.005	0.005	0.887	2.048	0.433	0.227
Australia	13	1	2	10	8	0	63	13	52	0.444	0.028	0.006	0.006	0.887	2.000	0.444	0.273
Finland	12	1	2	8	6	2	63	12	78	0.415	0.011	0.021	0.021	0.887	2.143	0.414	0.493
New Zealand	12	1	2	9	3	0	63	12	25	0.403	0.002	0.001	0.001	0.887	2.206	0.402	0.314
Spain	11	1	3	8	6	0	63	11	36	0.386	0.028	0.000	0.000	0.887	2.302	0.386	0.246
Poland	11	1	2	8	10	0	63	11	30	0.397	0.005	0.006	0.006	0.887	2.238	0.396	0.576
Sweden	11	1	2	9	3	0	63	11	25	0.409	0.003	0.001	0.001	0.887	2.175	0.408	0.360
Norway	11	1	2	9	3	0	63	11	21	0.389	0.001	0.001	0.001	0.887	2.286	0.388	0.370

Label	Input degree	Periodic strong components	All neighbors	All core partitions	Multi-level Louvain communities	Hubs and authorities	Size of all domains	All degree	All weighted degrees	All closeness centralities	Between-ness centrality	Hub weights	Authority weights	Normal-ized size	Average distance from all domains	All proximity prestige values	Aggregate constraint
Czech Republic	11	1	2	9	3	0	63	11	20	0.406	0.001	0.002	0.002	0.887	2.190	0.405	0.385
South Africa	11	1	3	8	6	0	63	11	20	0.352	0.025	0.000	0.000	0.887	2.524	0.352	0.318
Belgium	10	1	3	8	4	0	63	10	141	0.384	0.026	0.000	0.000	0.887	2.317	0.383	0.416
Canada	10	1	2	8	4	2	63	10	123	0.394	0.026	0.022	0.022	0.887	2.254	0.394	0.373

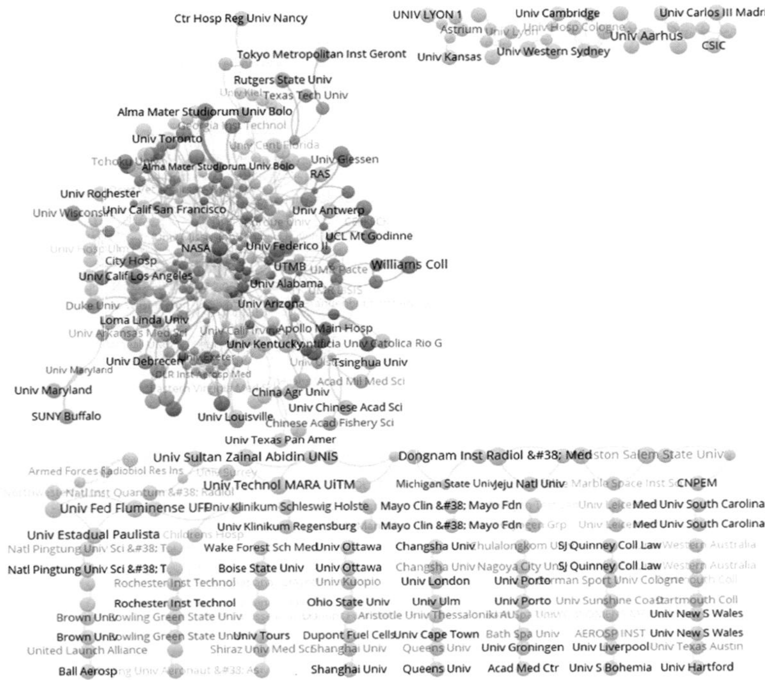

Figure 3.14 Institution collaboration network (degree)

the framework of technology and engineering management. In this respect, it can be stated that SNA is an important tool that should be preferred in TM. On the other hand, the use of mixed methods can be suggested for the detection of weak signals, visualization of the main competence points, and determination of the technologies already produced and the methods used in the development of these technologies. For this purpose, it is possible to use methods such as scientometrics, bibliometrics, SNA, graph theory, and ML.

Table 3.13 *Institutions network parameters*

Label	All-degree partition	Periodic strong components	Multi-level VOS clustering	Hubs and authorities	Size of all domains	All degrees	All weighted degrees	All closeness centralities	Betweenness centrality	Hub weights	Authority weights	Normalized size of all domains	Average distance from all domains	All proximity prestige values	All Laplacian centralities	Aggregate constraint
NASA	97	1	8	0	336	97	188	0.305	0.327	0.001	0.001	0.703	2.307	0.305	1 1310	0.045
University of California, Berkeley	28	1	40	0	336	28	35	0.249	0.075	0.001	0.001	0.703	2.824	0.249	1 700	0.104
Russian Academy of Sciences	25	1	28	0	336	25	37	0.230	0.056	0.000	0.000	0.703	3.060	0.230	1 468	0.263
Loma Linda University	24	1	10	0	336	24	89	0.215	0.024	0.000	0.000	0.703	3.268	0.215	1 070	0.148
Stanford University	24	1	19	0	336	24	36	0.233	0.026	0.000	0.000	0.703	3.021	0.233	1 246	0.157
Istituto Nazionale di Fisica Nucleare	24	1	34	0	336	24	31	0.204	0.026	0.000	0.000	0.703	3.452	0.204	1 002	0.194
Tohoku University	22	1	2	2	336	22	132	0.185	0.010	0.698	0.698	0.703	3.813	0.184	916	0.278
University of California, San Francisco	21	1	4	0	336	21	93	0.240	0.028	0.001	0.001	0.703	2.929	0.240	1 252	0.272
Virginia Commonwealth University	21	1	44	0	336	21	40	0.220	0.042	0.000	0.000	0.703	3.202	0.220	1 274	0.208
Baylor College of Medicine	19	1	9	0	336	19	42	0.232	0.013	0.001	0.001	0.703	3.030	0.232	1 072	0.316

Label	All-degree partition	Periodic strong components	Multi-level VOS clustering	Hubs and authorities	Size of all domains	All degrees	All weighted degrees	All closeness centralities	Betweenness centrality	Hub weights	Authority weights	Normalized size of all domains	Average distance from all domains	All proximity prestige values	All Laplacian centralities	Aggregate constraint
University of California, San Diego	19	1	66	0	336	19	25	0.226	0.037	0.000	0.000	0.703	3.107	0.226	854	0.162
Gunma University	18	1	52	2	336	18	26	0.188	0.006	0.059	0.059	0.703	3.750	0.187	766	0.260
University of California Irvine	16	1	14	0	336	16	43	0.242	0.030	0.000	0.000	0.703	2.905	0.242	822	0.152
University of Tokyo	16	1	51	2	336	16	22	0.193	0.016	0.054	0.054	0.703	3.640	0.193	614	0.270
University of Tsukuba	15	1	2	2	336	15	47	0.182	0.002	0.291	0.291	0.703	3.857	0.182	582	0.578
University of Antwerp	15	1	6	0	336	15	69	0.184	0.017	0.000	0.000	0.703	3.818	0.184	510	0.257
Hunter Holmes McGuire VA Medical Center	15	1	16	0	336	15	56	0.169	0.000	0.000	0.000	0.703	4.164	0.169	632	0.282
Penn State University	15	1	16	0	336	15	56	0.169	0.000	0.000	0.000	0.703	4.164	0.169	632	0.282
University of Central Florida	15	1	36	0	336	15	24	0.190	0.014	0.000	0.000	0.703	3.699	0.190	452	0.195
University of Nevada	15	1	41	0	336	15	22	0.240	0.015	0.000	0.000	0.703	2.938	0.239	758	0.170

4. Integrated analyses

In this part of the study, the methods used in the analyses of each technology field in the previous sections are combined and exemplified. In this context, inferences were made about descriptive information for R&D activities with bibliometrics, and the collaborations of actors in R&D activities and technology areas and classes that gained importance were determined with graph theory. In addition, thanks to the analysis conducted on patent data, their use in creating competitive advantages is exemplified.

DIGITAL TWIN: OPPORTUNITIES AND THREADS FOR R&D MANAGEMENT

Introduction

A digital twin is defined as a digital copy of a living or virtual physical entity (Wikipedia, 2019). Companies have made great efforts to reduce their production costs by developing a method to be used in innovation management that involves turning the developments in information and communication technologies (ICTs) into advantages (Machado & Davim, 2020). Thanks to this method, a digital twin can be used to review the material to be produced with all the details before it is presented for the final use, and the error rate can be reduced to zero (Puig & Duran, 2010). In this respect, digital twin technology, which has very important advantages in avoiding the possible problems involved in the infancy of production processes, has become an important R&D instrument that is used starting from the development stages of many products and technologies (Delbrugger et al., 2017; Rodic, 2017). In this part of the study, which focused on the potential of the digital twin concept and its technologies, a series of analyses were carried out on the bibliographic data of the scientific studies included in the international literature. Utilizing the analysis results, the opportunities and threats for R&D management, which hold significant potential for R&D management, are addressed from various perspectives, such as the requirements of the companies in the context of digital transformation, technology infrastructure requirements, necessities to be taken into consideration in the context of information and knowledge management, individual competencies, and the content of the training given, especially engineering. The digital twin concept has started to be discussed in both academia and industry (Koren et al., 1993; Renaudin et al., 1994).

It can be said that the advantages in the application and development stages of digital twins, with their unique features, make them stand out as a technology that will have an impact on all processes involved in R&D activities, such as ISO 10303-239 product development and life-cycle standards. In this regard, understanding the dynamics of digital twins and the advantages and threats they pose will benefit decision makers in the administrative processes. Digital twins, which enable the real world to be simulated, use the data provided by sensors as inputs (Canedo, 2016). In this respect, the real-time data flow is very important for digital twins. In order to ensure proper data flow, a network of sensors and a cloud structure that controls this network must be established. Even with this structure, it should be noted that an important infrastructure change is required for R&D companies and organizations (DebRoy et al., 2017; Heinzerling et al., 2017; Schuh & Blum, 2016). In other words, digital transformation seems inevitable for companies wishing to benefit from the advantages of digital twins, of which there are many, especially in R&D processes. In our analysis, we focused on scientific research and patent intelligence about digital twins and the technologies studied were examined closely.

Data and Method

An online search was carried out in order to obtain the bibliographic data of international scientific studies published in WoS on digital twins. A total of 811 publications on digital twins were identified. The obtained bibliographic data were made ready for analysis and recorded in a relational database. The basic bibliometric indicators about the data were as follows: the timespan was from 1973 to 2021, the number of publications with one author was 60, and the collaboration index was calculated as 3.63. This high collaboration index shows that the production rate of studies with multiple authors is quite high. In this part of the study, the aim was to obtain data about the technology domain with the themes examined primarily with bibliometric analysis.

Keyword Analysis

The keyword analysis showed that the frequently used keywords in digital twin research relate to the concepts of multidisciplinary studies. This can be used as a result that gives us clues about the widespread use of digital twin technology. When we look closely at the nodes that play important roles in the network in terms of centrality values, it can be seen that the digital twin technologies and methods have very serious potential for all production processes. In this respect, it is possible to say that the digital twin can be considered as a technology that shows emerging technology features. This situation is similar to the results presented in previous studies in the literature (Ameri & Sabbagh, 2016; Canedo, 2016; Schroeder et al., 2016). When we examine the concepts used closely, it can be seen that concepts that prioritize method, such as design, prediction, and genetic algorithm, come to the fore, and it is possible to state that this is evidence showing that digital twin research is in an early stage (Figure 4.1). It is a technology that actors aim to use because of its very high potential, and the search for improvement in the technology manifests itself in the keywords used in research. This can be understood as a result of the fact that digital twins' usage area is spread over a very wide range, which therefore entails a desire for customization according to the characteristics of each technology domain.

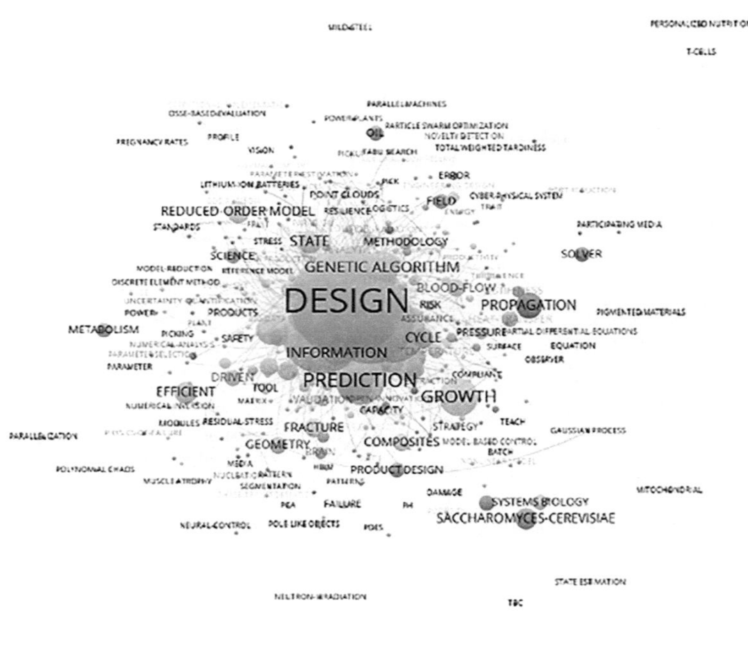

Figure 4.1 *Keyword analysis*

In order to identify the actors in digital twin research, a series of analyses were made based on author information. In these analyzes, the number of publications belonging to the authors were handled together with the number of citations, and the intersection points of the numbers of publications and citation numbers were tabulated together with the h-index information. In this respect, the tables below are provided to enable a holistic perspective in the examination of author performances.

Bibliometric Analysis

Author productivity

In the author productivity analysis, the authors' names were made uniform, and the performance indicators of the authors were calculated. Accordingly, the 20 most productive authors in digital twin studies are presented in Table 4.1.

Table 4.1 Author productivity

No.	Author	h-index	h-core citation sum	All citations	All articles
1	Tao, Fei	6	762	762	6
2	Leng, Jiewu	5	103	108	7
3	Liu, Qiang	5	163	178	9
4	Xu, Xun	5	234	234	6
5	Chen, Xin	5	207	212	7
6	Warmefjord, Kristina	5	128	130	9
7	Yan, Douxi	4	84	84	5
8	Nee, A. Y. C.	4	266	266	4
9	Park, Kyu Tae	4	35	37	6
10	Zhang, Ding	4	134	134	5
11	Tao, Fei	4	213	213	4
12	Soderberg, Rikard	4	52	61	10
13	Zhang, Meng	4	630	630	4
14	Lindkvist, Lars	4	121	121	6
15	Qi, Qinglin	3	446	446	3
16	Noh, Sang Do	3	26	28	5
17	Debroy, T.	3	106	106	4
18	Jiang, Pingyu	3	35	36	4
19	Barandiaran, Inigo	3	46	46	3
20	Chen, Chun-Hsien	3	102	106	6

It is seen that authors of Asian origin are at the top of the list in terms of author productivity. On the other hand, although Rikard Soderberg has the highest value in terms of the number of publications, he is in the middle of the list in terms of citation performance. For this reason, it is necessary to use similar

Table 4.2 Institution productivity

No.	Institution	h-index	h-core citation sum	All citations	All articles
1	Beihang University	11	1 014	1 037	21
2	Guangdong University of Technology	11	393	422	17
3	National University of Singapore	7	304	312	10
4	Shandong University	7	106	106	7
5	Penn State University	6	231	231	10
6	Chalmers University of Technology	6	161	175	15
7	Nanyang Technological University	6	204	216	14
8	University of Auckland	6	286	286	9
9	University of Cambridge	6	51	69	17
10	Hong Kong Polytechnic University	5	119	124	11
11	Texas A&M University	5	33	47	11
12	City University of Hong Kong	5	73	83	7
13	Chung-Ang University	5	28	28	6
14	Polytechnic University of Milan	5	102	107	12
15	Xi'an Jiaotong University	5	130	142	11
16	University of Sheffield	4	44	48	11
17	Northwestern Polytechnical University	4	50	53	7
18	Norwegian University of Science and Technology	4	46	48	10
19	Jiangsu University of Science and Technology	4	66	70	7
20	Sungkyunkwan University	4	35	43	11

methods that allow a holistic approach rather than depending on a single indicator in the analysis of author performance. This method can be seen as a tool that can be used for planning purposes in activities such as know-how transfer, invited speeches, training programs, and R&D budget allocation by using the field as a list of the most efficient authors in terms of both the numbers of publications and the impact they have made in the field.

Institution productivity

By using the affiliation information included in the author address information, the efficiency of the institutions operating in digital twin research was closely examined. As in the author performance analysis, the prominent institutions were handled together with the numbers of publications and citation numbers and tabulated with the h-index information expressing the intersection points of the numbers of publications and citation numbers. Accordingly, the top 20 institutions that stand out are given in Table 4.2.

In the institutional productivity analysis, it was observed that the institutions of Asian origin are at the top of the list. The representatives of the US, Penn State University, and New Zealand, the University of Auckland, can be seen as the other notable institutes. In addition to the list of researchers, the institution list is one of the tools that can be used in determining institutional expertise, especially in critical technology domains. The relevant list can be matched with the technology domain competencies obtained in patent analysis and used in determining parameters such as technological knowledge status and technological knowledge redundancy. In the table on productivity analysis in terms of the countries where institutions are located, the effect of the US is clearly seen, unlike the effects of individual efforts in the author and institution performance tables. Although it can be observed that China has better performance than the US in terms of the number of publications and the number of citations, the US ranks first in terms of h-index values (Table 4.3). When we look at the list in terms of countries, it can be observed that industrialized countries take their place in the list. It is not surprising that digital twin technologies, which offer significant advantages in meeting one of the most basic requirements of Industry 4.0, are being studied intensively in these countries.

Social Network Analysis

Regardless of the number of publications and citation indicators, the actors in the digital twin research network were closely examined with SNA. Author–author, institution–institution, and country–country relations were identified to determine the roles they play in the network. By using the centrality values to create network maps, the actors in the network were arranged according to their importance levels.

Author collaboration networks

In terms of author productivity, it is worth noting that more than one author cluster, albeit small-scale, appeared in the measurement results based on SNA. In particular, it can clearly be seen that author collaboration models are separated from each other in different clusters (Figure 4.2).

Although it was previously reported that the cooperation index was above three, the high number of clusters emerging in terms of network values can be interpreted as illustrating the fact that these clusters are located away from the center and that the necessary critical density in digital twin studies has not yet been reached.

Table 4.3 Country productivity

No.	Country	h-index	h-core citation sum	All citations	All articles
1	United States	26	1 190	1 761	239
2	People's Republic of China	22	1 656	3 118	352
3	Singapore	12	500	532	27
4	Italy	12	226	425	89
5	United Kingdom	11	276	678	153
6	France	11	443	522	69
7	Spain	10	176	246	69
8	Germany	10	467	750	225
9	Sweden	9	359	435	41
10	South Korea	9	125	259	71
11	Australia	7	203	255	37
12	New Zealand	7	301	301	11
13	South Africa	6	46	47	7
14	India	6	117	182	41
15	Denmark	6	76	81	20
16	Norway	5	64	88	30
17	Portugal	5	306	317	14
18	Japan	5	48	63	20
19	Netherlands	5	65	107	35
20	Finland	5	67	75	24

Institution collaboration analysis

In the institutional collaboration analysis, the address information of the authors was used. A situation similar as the one in the author collaboration map emerged in the institutional cooperation analysis (Figure 4.3).

In this analysis conducted based on institution addresses, it can be seen that important institution names such as Wuhan University of Technology, Chalmers University of Technology, Polytechnic University of Milan, Tampere University, and Stanford University stand out. The figure shows that there are nine main clusters based on corporate cooperation. In this respect, it is possible to say that digital twin applications and research have a multidisciplinary structure and the cooperation pattern is quite high, as stated in the introduction.

Country collaboration analysis

Looking the country collaboration models, it is worth noting that joint works are primarily conducted within institutions and therefore within countries. However, it is possible to see that there are studies in which countries work

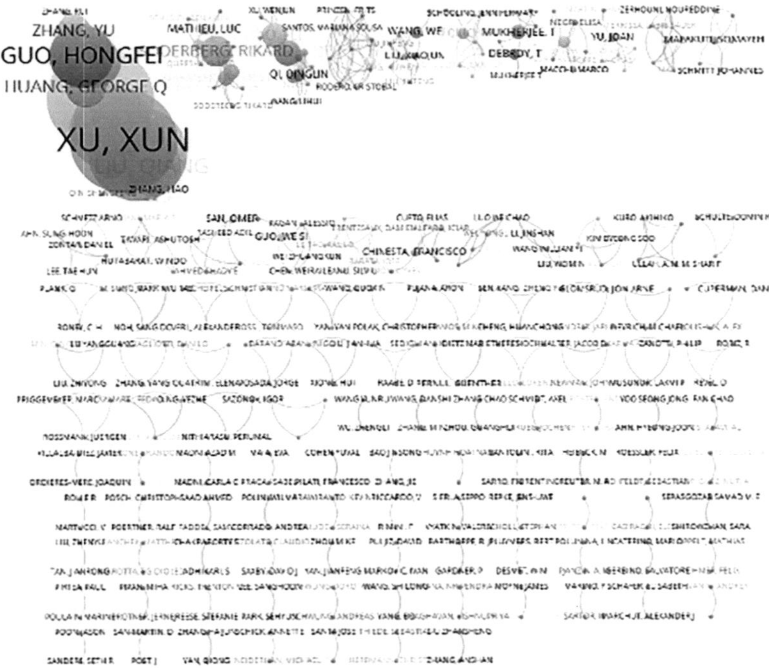

Figure 4.2 *Author collaboration network*

together in terms of international collaboration. The UK, the People's Republic of China, Spain, France, and Norway appear to be at the center of the digital twin studies network in terms of centrality values. Mexico, Serbia, Greece, and Ukraine also operate within their own networks outside of the network of digital twin studies (Figure 4.4).

Patent Analysis

Digital twin technology is a converging technology that involves producing a digital copy of physical or virtual assets. In addition to performing analyses through digital simulation and modeling real-time data flows, digital twins have high potential for the minimization of errors in product life-cycle management related to aircraft manufacturing, human–machine collaboration, and general manufacturing. In this part of the study, the key technology domains where most of the digital twin patents are located were explored and the distri-

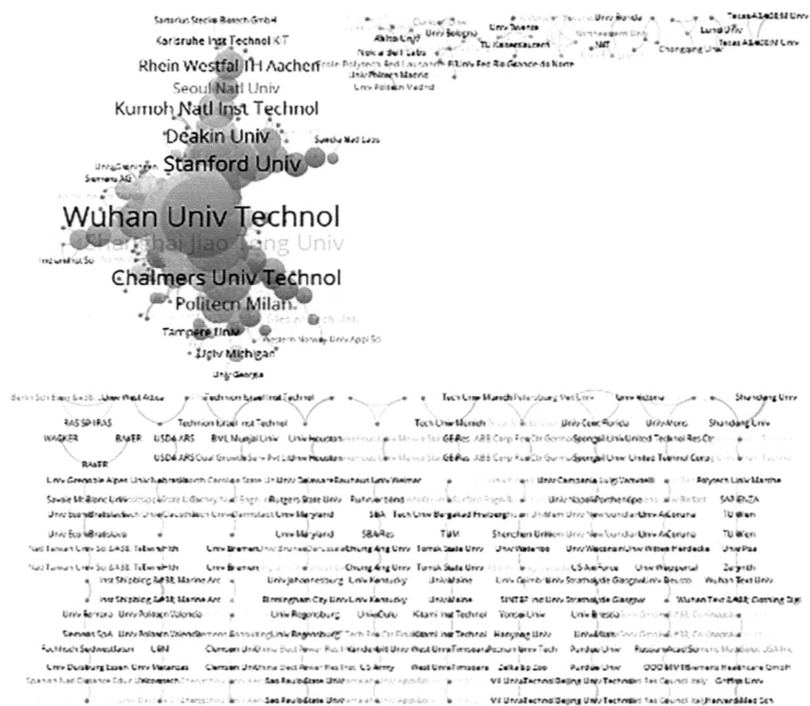

Figure 4.3 Institutional collaboration network

butions of the key technology domains, patent owners in the field, and patent placement strategies were determined based on SNA.

The first patent analysis conducted was an analysis of patent application ownership. With this information, it was possible to make inferences about short-, medium-, and long-term technology strategy development based on the outputs of patent intelligence analysis. In this context, an online search was made in the Lens database to access bibliographic data on patent applications for digital twin technologies. A total of 2032 patent documents were obtained.

Patent assignee analysis

Although there are several types of analyses that can be undertaken in patent analysis, we identified patent application ownership and related technology classes or domains. Looking at the literature, it can be seen that patents are evaluated against the technology domains they are related to (Wang et al., 2020); however, in our study, we identified technologies that are open to development

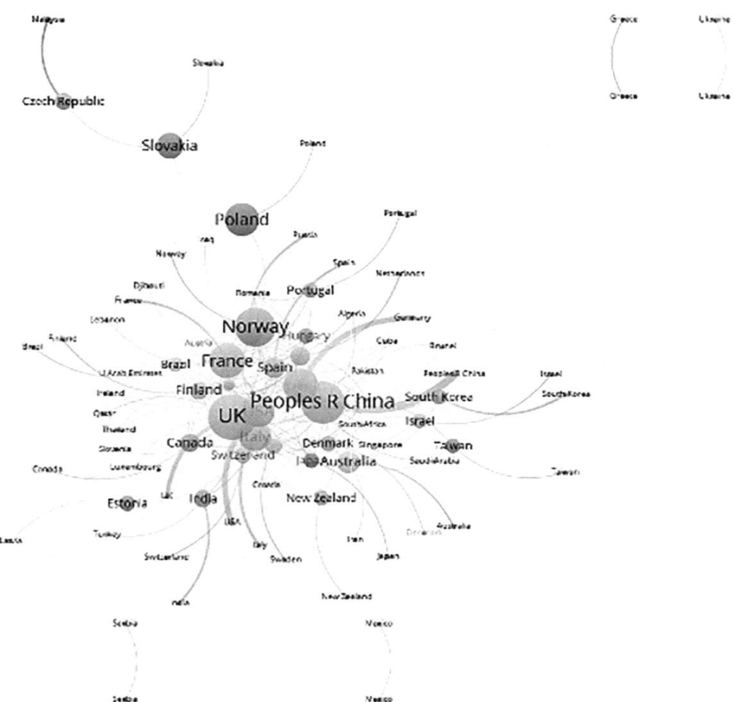

Figure 4.4 *Country collaboration network*

with a more holistic perspective in terms of strategy development. A series of analyses were carried out on the applicant ownership information for the 2032 patent documents. Accordingly, the details of patent application ownership are presented in Table 4.4. As can be clearly seen in the table, patent applicants for digital twin technology consist of representatives from almost every sector. In this respect, it is possible to say that digital twin technology has a wide range of uses. Although digital twin technologies were originally introduced in relation to simulation programs developed for NASA's matching technology (Glaessgen & Stargel, 2012), they have been shaped by industry giants such as IBM, Microsoft, Tesla, and General Electric. General Electric uses digital twin applications in aircraft engine production. When conducting tests of the aircraft engine in a physical environment, the data obtained in real time with digital twin technology is used to determine possible failure situations, the material aging time, and other conditions in order to decide whether the engine can be renewed or not. In the automotive sector, Tesla uses digital twin technologies to determine the engine life of vehicles, the mechanical aging of parts, damage situations that may occur in possible accident scenarios, and errors related to aerodynamic design. It is worth noting that General Electric, for example, reports that digital twins have saved their customers USD 1.5 billion thanks to their real-time data-based capabilities.

Table 4.4 Patent assignees and sectors

No.	Assignee	Sector	No.
1	General Electric	Energy, technology, and finance	323
2	Siemens AG	Industry, energy, and healthcare	165
3	Desktop Metal Incorporated	3D printing	107
4	Strong Force TX Portfolio	Innovation portfolio	95
5	Pure Storage Incorporated	Storage	85
6	Johnson Controls	Fire, HVAC, and security equipment	75
7	Honeywell International Incorporated	Aerospace, automation control, and energy efficiency	34
8	Microsoft Technology Licensing LLC	Technology	32
9	Beihang University	University	27
10	Rockwell Automation Tech Incorporated	Smart manufacturing	22
11	ABB Schweiz AG	Electrification, robotics, automation, and motion portfolio	21
12	Guangdong University of Technology	University	20
13	nChain Holdings Ltd	Blockchain	18
14	Accenture Global Solutions Ltd	Technology	16
15	Siemens Corporation	Industry, energy, and healthcare	16
16	SAP SE	Software	15
17	Myomega Systems GmbH	Internet of Things	12
18	Bentley Systems Incorporated	Software	12
19	Tata Consultancy Services Ltd	Technology and consultancy	12
20	PTC Incorporated	Computers and software	12

In the SNA carried out, when the network created by the patent applicants was examined, it was observed that several clusters have developed for digital twins. The map in which the network elements are positioned according to their centrality values is presented in Figure 4.5. It can be seen that the results are similar to those at the top of the patent application table, that is, the values for the assignee ownership.

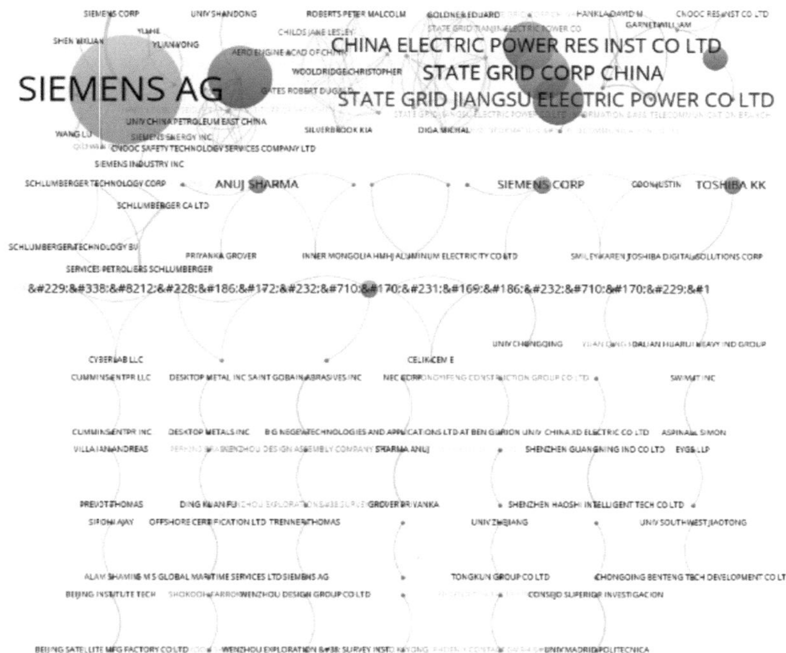

Figure 4.5 *Patent applicants network*

Technology Domain Network Analysis

SNA was conducted to identify key areas where digital twin patents concentrate, determine the relationships between key technologies, and identify technologies open to development. The network map that emerged as a result of the analysis enabled us to visualize the technology domains or classes that have reached the maturity level among digital twin technologies. Another important advantage is that it enables the creation of a table which is open to improvement, that is, it is possible to use it to determine further, new technology classes (Figure 4.6).

In terms of centrality values, it can be observed that the G05B19/418 technology class plays an important role for the digital twin patents network. When the related patent classes are examined closely, it can be observed that they are included in the field in which sub-technology domains for unmanned production activities, flexible production systems, and integrated production systems are classified. In this respect, it is possible to say that the minimization

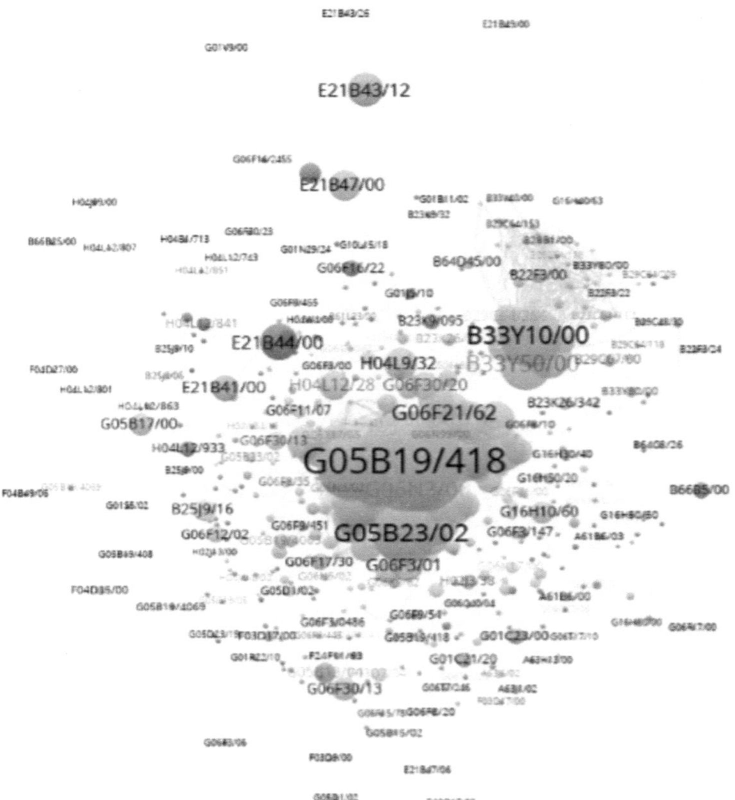

Figure 4.6 Technology domain network

of errors prior to the production phase through real-time data obtained from sensor networks is intensively examined in digital twin patenting activities. Structural gap analysis was applied to determine the technology classes reaching saturation level and those open to improvement (Table 4.5). Accordingly, it can be said that technology classes with high aggregate constraints have a higher maturity level for digital twin technologies than technology classes with low aggregate constraints. In other words, the technology class codes in the LAC header given in Table 4.5 can be examined to determine the technology classes that are open to development among digital twin technologies.

As a virtual model of a product, process, or service, creating a digital twin means creating the exact equivalent of the physical thing, a virtual twin. Digital

Table 4.5 Top 20 digital twin technology domains in terms of betweenness centralities, degree, and structural holes

HAC	LAC	Betweenness	Degree
G01B11/24	H04L29/08	G05B19/418	G05B19/418
B23Q17/09	G06F17/50	G06Q10/06	G06Q10/06
G05D7/06	G05B19/042	H04L29/08	G06N20/00
G06F16/2455	G06F30/20	G06F30/20	H04L29/08
G06Q99/00	H04L29/06	G05B23/02	H04L29/06
G05D23/19	H04L29/06	H04L29/08	G06N3/08
H04B1/713	G06K9/00	H04L29/06	G06K9/62
H04J99/00	G06F3/06	B33Y10/00	G05B23/02
G16H40/63	G06N20/00	G06N20/00	G06N5/04
G16H20/17	G05B15/02	G06K9/00	H04L29/08
G06F13/40	H04L9/06	G06N3/08	G06N3/04
G06F11/14	G05B13/04	B33Y50/00	G06Q50/06
G06F30/23	G06F21/62	B22F3/105	G06F30/20
G06F15/78	G06Q10/00	G06F21/62	G06Q30/02
G06F30/00	G06N99/00	H04L29/06	G06Q10/04
H02J7/00	G06T17/00	B64C39/02	B33Y10/00
B23Q17/09	G06F8/60	G06F3/06	G06F16/23
G05B19/4069	G06T19/00	G05B15/02	G06K9/00
B66B25/00	G05B19/418	G06F17/50	B22F3/105
G05D1/02	G06T19/00	B33Y30/00	G06F9/50

twins are virtual replicas that data analysts and IT professionals can simulate before building real devices. Digital twins are not only used in production but also as a tool that has an impact on the development of technologies such as the IoT, AI, and data analytics. It is possible to say that the concept of the digital twin, which helps IT professionals and data analysts achieve high efficiency and an optimal distribution of resources, is now spreading across a wide spectrum, from the aircraft industry to the agricultural sector.

Thanks to this case study, we were able to closely examine the ways in which digital twins, which have such a wide range of potential uses, have been taken up in both R&D studies and in the dynamics of commercialization by being patented in the context of original ideas. From this point of view, the sampling of the analyses carried out reveals that such results can give important clues to decision makers in terms of strategy development. It is hoped that this guide will be used as an integral tool in technology and engineering management that can be used in policy and strategy development.

AN ANALYSIS OF SUPERCOMPUTING TECHNOLOGIES: AN APPLICATION BASED ON BUSINESS INTELLIGENCE

The term 'supercomputer' was first used in the New York World newspaper in 1929. It is known that the first specially designed superior tabulator computer was built by IBM for Columbia University in 1931 (Borbély, 2011). Supercomputing is the process carried out by dense parallel processors, high-performance vector processors, and systems formed by cluster computers. In scientific activities it enables the formation of new methodologies, such as computational science, that combine scientific processes with computing (Rivas et al., 2017). Supercomputing operations carried out with computers that are much more powerful and of higher capacity than general-purpose personal computers are used extensively for very large volumes of data, solving very complex problems, and policy activities that require the use of large amounts of data (Lang et al., 1995).

Developments in information and communication technologies have led to the emergence of very large datasets and increased the need to structure these data. In this respect, the need for computers with high computing and processing capacities has increased. The requirements for the execution of high-capacity operations have also changed in this context, requiring all equipment, from database management systems to processor architectures, to be considered within the scope of supercomputing (Blumrich et al., 2009, 2010; Pietrzyk et al., 2019). Research on the subject has started to increase day by day and international journals that publish directly in the field of supercomputing have emerged. Supercomputing, which can be witnessed in many subfields, notably AI and medical imaging technologies, is especially extensively used in neuroscience (Witten, 1991). In this respect, the application area is quite wide and, at the same time, supercomputing has to be handled as an emerging technology in all aspects. The leading actors in terms of supercomputing authors, institutions, and funders were identified using a bibliometric analysis (Yalçin & Şeker, 2020). SNA was used to identify collaborations and, in this context, author–author, institution–institution, and country–country cooperation networks were created. Finally, patent analysis was conducted in order to analyze the information for the registered ideas about supercomputing technology. Using the obtained data, the supercomputing technologies and sub-technologies, if any, that have reached the saturation level and which sub-technologies are open to development were determined. In this part of the study, a TM setup is exemplified by using mixed methods with a holistic perspective.

INTRODUCTION

Technology management has recently been emerging as a very important concept. A wide range of technologies, from energy engineering to educational technologies, can make use of it. However, the increasing amount of information, especially due to the developments in information and communication technologies, brings along structured and unstructured data approaches; the management of untouched data especially has turned into a very important challenge for information managers. The TM method found in the technology management literature can be used to solve this problem. Thanks to this method, technology management principles can be combined with bibliometric and scientometric methods (Bakhtin & Saritas, 2016; Cho & Daim, 2016; Daim et al., 2016; Madani & Weber, 2016; Martin & Daim, 2012). Furthermore, by making use of statistical power laws, a meaningful decision-support mechanism can be used in the decision-making process for the development of instruments. When we look at the literature, bibliometrics, scientometrics, patent analysis, and business intelligence can be seen among the preferred analysis methods in technology management (Madani & Weber, 2016). Supercomputing can be seen to be among the important methods and practices in this context. It is possible to identify a series of studies in the literature regarding this technology, and we were able to fetch about 1400 scientific articles WoS. From this point of view, it is possible to say that supercomputing is an emerging technology field. In order to be able to examine this technology field with all its components, it should be considered from a holistic perspective. In this context, the main actors were determined with a bibliometric analysis to reveal their basic characteristics. SNA and graph theory were used in order to determine patterns of collaboration, and maps based on the SNA of author–author, institution–institution, and country–country collaborations are provided.

A patent analysis was also carried out to determine the application statuses of the ideas in the field of supercomputing technologies and to determine the applicants and technology owners with the most patents in this technology domain. The patent analysis focused especially on technology classes. In this context, the technologies that reached a saturation point were detected and structural gap analysis was used to determine the technology classes that are open to improvement. In other words, it was possible to identify sub-technology domains, which we can call open-to-development areas. However, SNA was used to reveal the relations of technology classes with each other. To do that, the centrality measurement parameters from social network theory were used. For the centrality degree, determinations were made regarding indicators such as centrality, creation of hubs, and prestige values. In this way, it was possible to visualize the relations of important technology classes in the supercomputer technology domain with other technologies on the social network map. The relationship sets that were visualized on these maps provided us with opportunities to identify the focal points and the relations to the outside. Thanks to these possibilities, we were able to draw a general profile of supercomputing technologies within the scope of TM. Thanks to the case study carried out, we identified in detail the degree to which applications of mixed systems, mixed methods, bibliometrics, scientometrics, SNA, and patent analysis can support decision makers in technology and engineering management.

In the next part of the study, descriptive statistics extracted from relevant databases for the bibliographic data relating to scientific research articles in the international literature are presented. Descriptive information about the field was fetched in this way and, in the SNA, examples of how these maps can be created by using open source and free software were produced. Information about the scales that are important for centrality values is also presented. This information is useful in determining the scale and purpose of information obtained. Identifying the authors who frequently carry out scientific research in the domain of supercomputing technologies, what subjects these authors work on, and which countries their institutions are located in can actually provide very useful information in determining the technological knowledge status (TKS) and the level of technology knowledge readiness (Daim et al., 2020). In determining the performance values for each author, outputs were obtained using basic bibliometric indicators, such as the number of publications of the authors, the number of citations they had received, the number of co-authorships, and the determination of authors' potential to be team leaders based on their status as first or correspondence authors. Thanks to these outputs, it was possible to identify the main actors related to the field. After the authors were identified, the necessary information was analyzed using the text mining method and the address information provided by the authors in their scientific research articles.

To determine the most productive institutions in the field, values such as the number of publications and the number of citations were examined. Matching matrices were created to calculate the h-index, which represents the intersection point of the numbers of publications and numbers in a country. Thanks to these matrices, it was possible not only to analyze performance based on the number of publications, but also to compare the effects of those publications in the literature. When we examined the studies within the scope of the patent analysis, bibliographic information of patent documents on supercomputing technologies was compiled from international patent databases such as the USPTO, EPO, Japan Patent Office (JPO), and WIPO. We stored the data that we fetched as a relational database for analysis. The database was cleared to make it ready for analysis. The classification numbers of each patent included in the data were put into a uniform format, both the classification numbers given by the international patent office (the International Patent Classifications (IPCs)) and the CPC codes. The patents were divided into technology classes and analyzed. In this way, the identification of patent applicants, who are intensely trying to patent in the field of supercomputing technologies, was achieved at the institutional level. The applicants who applied individually as inventors were determined. Considering the distribution of patent applications by year, the annual growth rate for the technology domain was calculated. This annual growth rate was evaluated within the framework of growth stages, and inferences were made based on the curve models that they conformed to using growth curve theories. In this way, it was possible to make inferences about whether supercomputing technologies had reached a certain saturation level at the point of the comparison of growth stages. Based on the information we obtained from the methods and theories used in technology mining, we inferred that supercomputing technology is an emerging technology area.

Taken within this framework, the analyses once again demonstrate that supercomputing technologies are of great importance in terms of adapting to today's changing and challenging conditions, as they allow calculations to be made for large amounts of data. The analyses made it possible to identify the nodes that have reached the saturation level for connectedness in terms of social network values and which are open to development, especially with the technology classes that reach saturation level in terms of the comparison of technology classes. In this respect, it is possible to say that the study conducted is a guide for decision makers, especially in the field of technology and engineering management. The sampling in this study also demonstrates the tool that can be used in identifying and tracking technologies that gain importance, which we call emerging technologies, and in determining short-, medium-, and long-term strategy policies for these technologies.

Results

Starting the analysis with the presentation of bibliometric indicators is important in terms of determining the fundamental features of the technology area. A total of 1410 publications were analyzed, and the annual growth rate for supercomputing studies was found to be 12.37 percent. This value indicates that the related technology will show an increasing trend in the coming years as a technology field. Furthermore, in the context of the query strategy, the data covered publications between 1978 and 2020. The number of publications with a single author was 369. Considering that the collaboration index was calculated as 3.89, it can be seen that co-authorship is dominant in the research in the field of supercomputing.

Identifying supercomputing research trends

To identify supercomputing research trends, a co-word analysis was carried out on the frequency that keywords were used together. In this way, it was possible to identify prominent research foci in the field, as well as to visualize the concepts for sub-technology areas addressed in the supercomputing domain. The sizes of all nodes in the grid map are represented as circles. The sizes of the circles and the sizes of the caps are proportional to the frequency values. Lines of relationships between concepts are shown. The thickness of these connections is proportional to the density of the relationship between concepts. The connectivity coefficient was used in determining the sets (Figure 4.7).

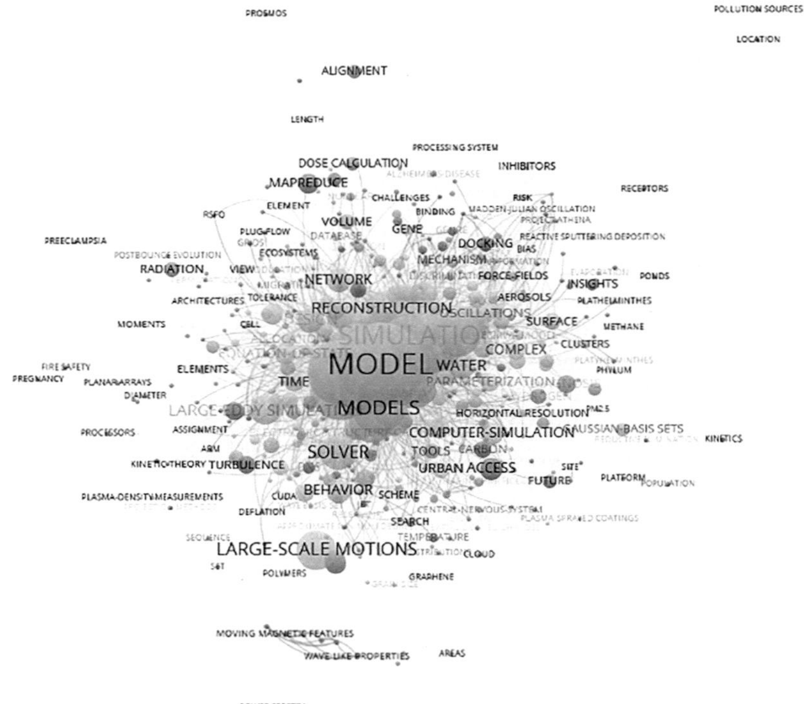

Figure 4.7 Research trends for supercomputing

Author Productivity Analysis

It is clear that the determination of the most productive authors, institutions, and countries in the field of supercomputing technologies can contribute to strategy formulation for decision makers in technology and engineering management. In this respect, to determine the most productive authors in related technology studies, performance indicators based on the number of publications were determined and then the h-index values that form the intersection of the number of publications and the number of citations were compared. Details on author productivity are presented in Table 4.6. This table is important as it shows who dominates the basic research about the technology domain. Although it can be seen that the authors perform similarly in terms of the number of publications, when considered in terms of citation values, it is clear that this table provides important information. Considering that supercomputing refers to the processing of highly complex or data-laden problems using the dedicated computing resources of multiple computer systems operating in parallel, it would be correct to say that supercomputing requires researchers from more than one discipline to work together in the technology domain. Supercomputing can be described as a domain of technology that produces solutions through problem solving and data analysis that are impossible to achieve with standard computers, being too time consuming or costly. The high-performance computing that supercomputing provides is very important as it can be used in decision-support mechanisms to produce meaningful information from big datasets. In this respect, it was an expected result that the basic research on this technology, which has a high potential of use in areas such as weather, energy, the life sciences, and production management, is multi-authored (Table 4.6).

Source Productivity Analysis

As an emerging technology, the publication of a dedicated journal on this subject can be seen as an important indicator since many subareas of supercomputing still need to be investigated for basic research. As well as the hardware side of supercomputing, other aspects such as software, interoperability, architecture and systems, algorithms, languages and programs, and performance measures are also important. From this point of view, it becomes very important to determine the scientific resources that are focused on research on the supercomputing technology domain. Utilizing the results of the analysis conducted for this purpose, the top 20 sources in which research is most frequently published are presented. The journals in which the authors frequently publish their publications can also be used to identify the core sources of the relevant technology. Thanks to this information, it is possible to identify the resources required to keep track of qualified information in the field. In this respect, the journals in which scientific studies on supercomputing technologies are frequently published were determined and presented along with their h-index values (Table 4.7), as in the author productivity table. It can

Table 4.6 Author productivity

No.	Author	h-index	h-core citation sum	All citations	All articles
1	Chrisochoides, Nikos P.	4	44	44	4
2	Aloisio, G.	4	72	72	4
3	Coveney, Peter V.	4	211	211	4
4	Xu, Ji	4	188	188	4
5	Ruede, Ulrich	4	89	92	5
6	Li, Jinghai	4	188	188	4
7	Kesselman, C.	3	68	68	3
8	Miyoshi, Takemasa	3	73	73	3
9	Wang, Limin	3	157	157	3
10	Ge, Wei	3	175	175	3
11	Frenkel, Yevgeniy	3	100	100	3
12	Franke, H.	3	146	146	3
13	Zenios, S.	3	45	45	3
14	Griffith, D.	3	35	35	3
15	Butler, D.	3	24	24	4
16	Moreira, J.	3	101	101	3
17	Zhang, Y.	3	101	101	3
18	Wang, F.	3	76	76	3
19	Cafaro, M.	3	36	36	3
20	Riedel, Morris	3	57	57	4

be seen that scientific studies on supercomputing technologies are frequently published in journals that also undertake their publication activities in the field of computer science. It can be observed that the supercomputing publications, which are at a limited level in terms of the number of publications, show noticeable performance in terms of numbers of citations even in the early period, and the top ten journals in the list have similar values in terms of their h-index values. According to Bradford's scatter analysis (Vickery, 1948), with which the scattering or distribution of the literature on a particular subject can be determined, 19 of the top 20 most productive sources for supercomputing technologies are in the first region. In other words, 19 out of 20 sources given in Table 4.7 constitute the core journal collection that publishes research on supercomputing technologies.

Country productivity

Another level to be considered in productivity analysis is that of countries and institutions. In this context, first of all, the countries leading in research

Table 4.7 *Source productivity analysis*

No.	Journal title	h-index	h-core citation sum	All citations	All articles
1	*Computer*	11	392	414	29
2	*Computer Physics Communications*	9	243	291	22
3	*Journal of Parallel and Distributed Computing*	9	152	188	21
4	*Future Generation Computer Systems – The International Journal of eScience*	8	141	165	19
5	*Parallel Computing*	8	248	310	34
6	*Proceedings of the IEEE*	8	340	346	11
7	*IEEE Transactions on Computers*	8	849	850	9
8	*IEEE Transactions on Parallel and Distributed Systems*	7	247	265	14
9	*Concurrency and Computation – Practice and Experience*	7	234	281	25
10	*International Journal of Supercomputer Applications and High-performance Computing*	6	108	131	26
11	*Computers and Geosciences*	6	161	161	6
12	*Journal of Supercomputing*	6	52	85	36
13	*Journal of Computational Physics*	6	284	284	6
14	*SIAM Journal on Scientific Computing*	6	173	182	9
15	*Computing in Science and Engineering*	5	69	76	13
16	*IEEE Transactions on Antennas and Propagation*	5	124	124	6
17	*Bulletin of the American Meteorological Society*	5	121	124	8
18	*Nature*	5	66	67	15
19	*Concurrency – Practice and Experience*	5	78	96	12
20	*Monthly Notices of the Royal Astronomical Society*	5	367	368	6

activities in supercomputing technologies were determined. The numbers of publications and the impact indicators of the countries are shown in Table 4.8. The US being at the top of the list is an expected result considering that it is the country where the first supercomputer was built. However, it can be seen that the countries in the top positions of the lists are EU member states, except for China, which is in sixth place, and Australia, which is ninth. It can be inferred that this results from the impact of the ERA-Net ERA-LEARN project carried out by the EU.

Table 4.8 Country productivity

No.	Country	h-index	h-core citation sum	All citations	All articles
1	United States	69	10 264	23 462	1 165
2	United Kingdom	29	2 308	3 168	150
3	Germany	29	2 740	3 921	170
4	Japan	27	1 526	2 085	109
5	Spain	25	2 538	3 341	143
6	People's Republic of China	23	1 102	1 705	185
7	Italy	20	924	1 300	78
8	France	19	2 408	2 616	63
9	Australia	17	1 041	1 422	97
10	Switzerland	13	362	539	56
11	South Korea	9	274	338	43
12	Russia	9	172	318	72
13	Austria	8	174	174	9
14	Canada	8	520	608	33
15	Netherlands	7	158	192	24
16	Saudi Arabia	7	147	155	13
17	India	7	144	168	29
18	Mexico	6	52	52	12
19	Taiwan	6	90	133	28
20	Poland	6	97	135	21

When the performances of the countries are examined, it can be seen that the US is at the top of the list in terms of numbers of publications and citations and h-index values. Although China ranks second in terms of publication performance, unlike the other technology fields we have examined within the scope of this study, it is lagging behind countries such as the United Kingdom, Germany, Japan, and Spain in terms of impact values.

Institution productivity

In terms of institution productivity, institutions with US addresses lead the list in terms of both the number of publications and their impact indicators. The contributions of two laboratories supported by the American Ministry of Energy attract attention, followed by the Chinese Academy of Sciences (Table 4.9).

Table 4.9 Institution productivity

No.	Institution	h-index	h-core citation sum	All citations	All articles
1	University of Illinois	19	1 114	1 335	68
2	Harvard University	14	1 529	1 532	17
3	University of California, Berkeley	13	844	942	33
4	University of California, San Diego	13	1 219	1 281	25
5	Oak Ridge National Laboratory*	12	383	500	45
6	Argonne National Laboratory	12	782	843	29
7	University of Tennessee**	12	373	450	29
8	Chinese Academy of Sciences	10	364	403	25
9	University of Maryland	10	418	456	16
10	NASA	10	240	282	23
11	Lawrence Berkeley National Laboratory	9	211	244	20
12	University of Southern California	9	421	433	11
13	University of Texas at Austin	9	371	386	13
14	Stanford University	9	270	270	14
15	Forschungszentrum Jülich	8	239	260	13
16	University of Pittsburgh	8	213	241	15
17	University College London	8	302	308	14
18	Cornell University	8	446	452	11
19	New York University	8	371	397	13
20	University of Chicago	8	502	528	15

Notes: *Oak Ridge National Laboratory is an American multiprogram science and technology national laboratory sponsored by the U.S. Department of Energy and administered, managed, and operated by UT–Battelle as a federally funded research and development center under a contract with the DOE (Wikipedia, 2021b).

**Argonne National Laboratory is a science and engineering research national laboratory operated by UChicago Argonne LLC for the United States Department of Energy (Wikipedia, 2021a).

Social Network Analysis

By conducting author–author, institution–institution, and country–country analyses of collaborate, stakeholders engaged in R&D activities together in supercomputing research were determined. Network metrics calculated with SNA gave us important information about the importance levels of the actors in the network and their roles.

Author collaboration

In the author collaboration analysis, the names of each author were made uniform and graph theory visualizations are provided to show a clean image. Figure 4.8 shows a network map in which the authors are positioned according to their centrality values. Accordingly, it is possible to say that researchers such as Owen, Herbert; Vasquez, Mariano; Valero, Mateo; Riedel, Morris; Vishwanath, Venkatram; and Liao, Wei-Keng stand out in terms of their centrality values.

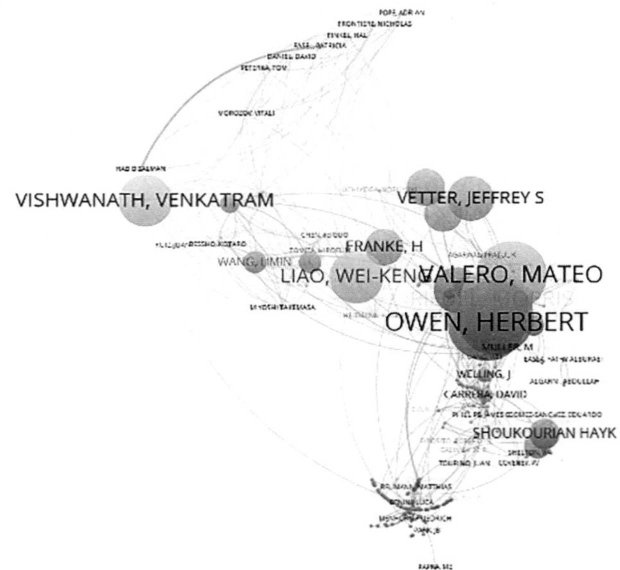

Figure 4.8 Author collaboration network

Institution collaboration

The network map is given in Figure 4.9 and allows us to examine the indicators based on the SNA of the research institutions that authors are affiliated to, emphasizing the importance of the network values and the roles they undertake. Considering the positions of these values and nodes in the network, it is notable that institutions such as the University of Tennessee; the University of Illinois; the University of California, Berkeley; the University of California, San Diego; Oak Ridge National Laboratory; the Chinese Academy of Sciences; the University of Maryland; Harvard University; Argonne National Laboratory; and Lawrence Berkeley National Laboratory seem to have taken on important roles. When this result is evaluated together with the table comparing the performances of the institutions, it is an important output that can be used in determining the dominant institutions in supercomputing research (Figure 4.9).

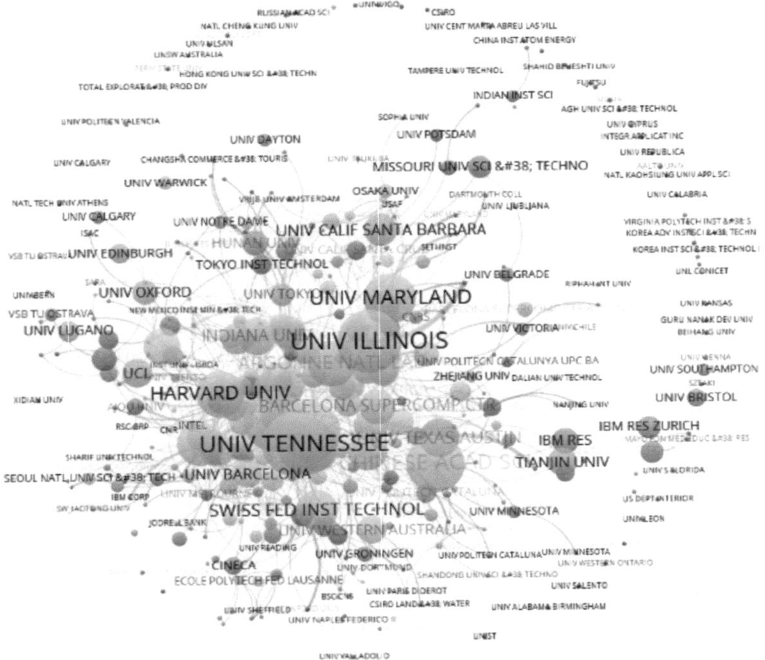

Figure 4.9 *Institution collaboration network*

Country collaboration

We also performed an SNA based on address information in order to visualize the position of countries in terms of their role in the network of cooperation dynamics and supercomputing research. As a result of the analysis carried out in this context, it was possible to determine the patterns of cooperation between countries and to see the dominant countries in the relevant research field. It should be noted that in the country cooperation analysis, a situation similar to the performance analysis of countries emerged. It was found that countries that stand out in terms of numbers of publications and citation values also stand out in terms of centrality values in the network. In this respect, the US, the UK, Spain, Germany, the People's Republic of China, France, Australia, Switzerland, Italy, and Japan, which are among the countries that carry out scientific studies on supercomputing technologies, appear to have a dominating effect (Figure 4.10).

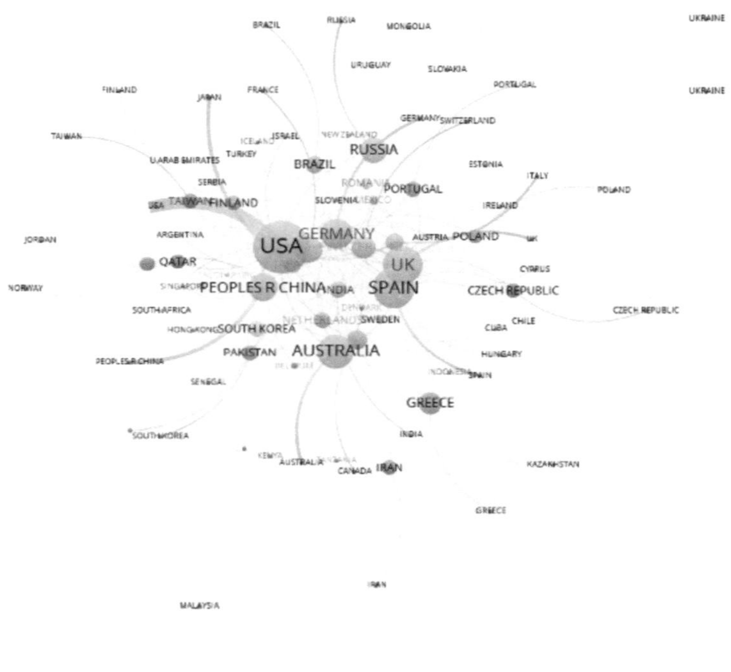

Figure 4.10 Country collaboration network

Patent Analysis

When we look closely at the patentability criteria, features such as innovation, invention step, and industrial applicability come to the foreground. Innovation means that the invention does not yet exist worldwide, in other words, it is not included in the state of the art. The state of the art, however, refers to all kinds of information (announced by written or oral promotion, usage, etc.) about the invention anywhere in the world before the patent application date. In the invention step, attention is paid to the fact that the invention cannot be evidently inferred from the state of the art by a person skilled in the related technical field. Industrial applicability means that the invention can be produced, applied, or used in any branch of industry, including agriculture (Abraham & Moitra, 2001; Daim et al., 2006; Madani & Weber, 2016; Tsenget al., 2007).

With the analyses conducted on supercomputing technologies through patent data, it is possible to address issues such as on the sub-technologies around which commercialized ideas in technology are concentrated and the institutions around which the patent ownership is concentrated,. In this context, according to our analysis of the dataset of 42 415 supercomputing patents that we compiled from Lens, we observed that the related patent literature could be explained by a linear curve from the growth curves ($R^2 = 0.875$) (Figure 4.11).

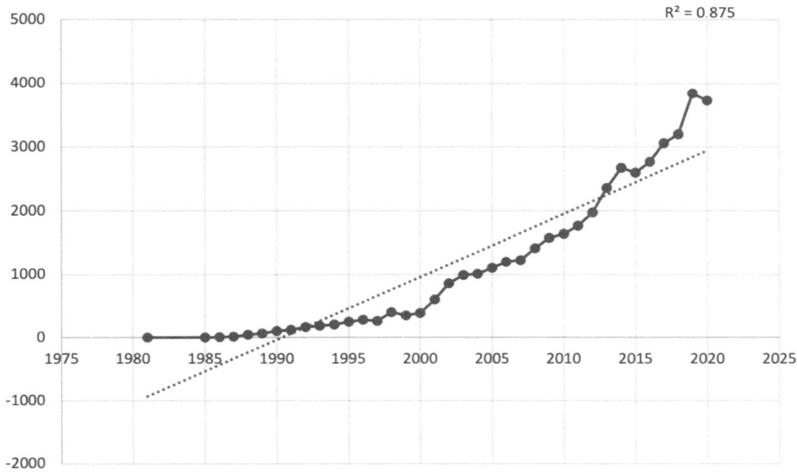

Figure 4.11 Supercomputing patent documents by year

When we consider the matter in terms of patent application ownership, IBM is at the top of the list, along with Intel, Rovi Guides, Microsoft, Microsoft Technology Licensing, Sun Microsystems, Oracle Corporation, Fujitsu, Facebook, Nvidia Corporation, and Intel (Table 4.10).

When evaluated in this respect, it can be seen that, among the technology giants, very strong companies in both software and hardware have made patent applications in the supercomputing domain. One of the striking results that stands out is Rovi Guides. Since Rovi Guides is active for the user experience

Table 4.10 Supercomputing patent appliers

Applier	No.
IBM	3 676
Intel	2 533
Rovi Guides	1 591
Microsoft	960
Microsoft Technology Licensing	530
Sun Microsystems	481
Oracle Corporation	473
Fujitsu	411
Facebook	385
Nvidia Corporation	357
Intel	313
United Video Properties	251
Apple	249
NEC Corporation	246
Digimarc Corporation	216
Intel IP Corporation	200
General Electric	198
Qualcomm	197
Amazon Technologies	193
Hewlett-Packard Development Company	189
Advanced Micro Devices	178

on the end-user side, it can be said that they play a role in the compilation and processing of end-user data. Similarly, the fact that Facebook is included in this list can be classed among the striking results for the supercomputing domain. With the structural gap analysis, technology classes that are open to improvement were determined and these indicators are presented in the table as HAC and LAC values. Accordingly, 'Video games' (A63F13/00) and 'Coupling light guides with opto-electronic elements' (G02B6/42) are the most open technology classes in the supercomputing technology domain. Table 4.11 can be examined for the technology classes formed in terms of details and centrality values.

SNA was conducted using patent classification codes in order to see the technology classes around which technology domains concentrate. Since the value of the difference between the degree of correlation and centrality is used in the placement of the network elements, the visualization of the technology classes that play important roles in a relevant technology domain is provided.

Table 4.11 Top 20 supercomputing technology domains in terms of betweenness centralities, degree, and structural holes

LAC	HAC	Degree (all)	Betweenness centrality	Weighted degree
A63F13/00	G06F17/30	G06F17/30	G06F17/30	G06F9/46
A63F9/24	G06Q10/00	H04L29/06	H04L29/06	H04L29/06
G02B6/42	G06F17/50	G06F15/16	G06F19/00	G06F17/30
G01V1/30	G06Q10/06	G06F13/00	G06F15/16	G06F13/00
G06F1/20	H04L29/06	G06F9/46	G06K9/00	G06T1/00
A61N1/372	G06F19/00	G06F15/173	G06F13/00	G07F17/16
G02F1/01	G06F21/10	H04L29/08	G06F17/50	G06F15/173
G06F7/00	G06F3/14	G06K9/00	G06F17/00	G06F12/00
G02B6/122	G06N20/00	G06F17/00	G06F9/46	G06F15/16
H01L21/56	G06F3/00	G06F12/00	G06F15/173	H04L1/00
G07F17/32	G06F13/40	G06F9/50	G06F9/44	H04N1/32
G01V1/28	H04L29/06	G06F9/44	G06F11/00	G07F7/08
H01L25/065	G06Q30/02	G06F3/00	H04L29/08	H04N1/00
H01L23/31	G06F17/00	G06F9/30	G06F3/00	G07F7/10
G02B6/43	G06F9/46	G06Q30/02	G06F12/00	G06F15/80
C12N15/82	G06Q50/00	G06F9/38	C12Q1/68	G07D7/00
H01L21/48	G06F11/34	G06F3/06	G06F9/50	G07D7/12
H04J14/02	G06F12/14	G06F11/30	G06K9/62	G06F12/08
G06F1/26	G06F13/42	G06F19/00	G06F11/30	G06K9/00
H05K3/46	G06F13/00	G06F9/455	G06F9/30	H04L29/08

When examined in this respect, it is possible to observe that the class of 'digital computing or data processing equipment or methods specially adapted for certain functions' is at the center of the network of inventions registered in the supercomputing field (Figure 4.12).

Supercomputing technologies have been of great importance throughout their history. These technologies, which have been used to make significant progress in scientific exploration and to deal with socially important problems, especially national security, are nowadays present in the fields of stock management, defense intelligence, climate forecasting and earthquake modeling, transportation, manufacturing, social health, and security. They are used to solve challenging problems in all areas. In order to understand such important technologies in all their aspects, it is obvious that, starting from the ways in which they are handled in basic research, examining them with methods based on product research and patent analysis can provide very important outputs for technology and engineering management.

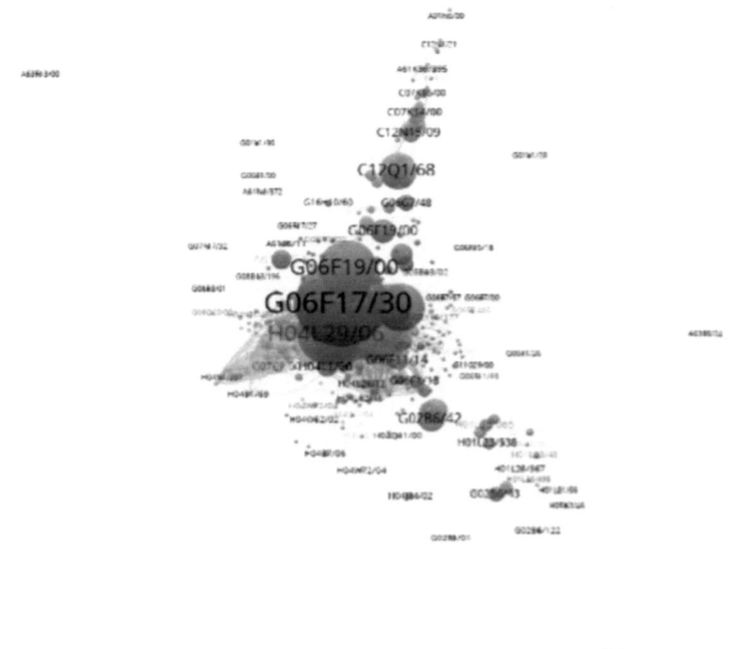

Figure 4.12 Technology domain network

In our study, we ensured that the main actors were identified at the individual, institutional, and country levels. Thanks to methods based on SNA, the roles of these actors in the technology domain could be classified according to their paying levels. In the patent analysis, the relationships between the technology domains were revealed, and the individuals and institutions that direct the technology were determined in terms of both patent ownership and inventors. It is thought that the methods and applications used in this study will be useful tools for field workers and researchers in technology and engineering management.

5. Conclusion

This book demonstrated the use of technology mining through an investigation of emerging technologies. The objectives of this book were multifold:

1. Provide a set of cases studies for academic institutions and students so that the book can be used in a class environment to teach technology mining;
2. Provide a baseline for graduate students and young researchers getting into this field;
3. Provide a set of guidelines for professionals in industry and government who are responsible for managing research and development so that they can use technology mining in their work.

We believe that we accomplished all these objectives throughout the book and hope that the book will add value to the existing knowledge.

As outlined at the introduction, our focus throughout the book was technologies that we believe are playing a great role in transforming our society. We decided on these technologies after reviewing lists published by different organizations and institutes, as well as from discussions with our colleagues.

The following table summarizes what is presented in the previous chapters. The results show that we were correct in selecting these technologies. The intelligence we gathered shows that these technologies demonstrate that leading governments, industries, and academic institutes around the world have been investing in and developing these technologies. We now briefly discuss each case based on the intelligence gathered regarding leading institutes and countries.

Let us first look at the medical transformation. In the case of COVID-19, we see a worldwide effort led by the US, China, and Germany. The leading research centers, on the other hand, are in the US and Hong Kong. In the case of AI and its use in the medical field, the US, China, and France are leading the effort. However, we see efforts concentrated in US institutes, including Harvard University, the University of North Carolina, and the Massachusetts Institute of Technology. The investigation of robotic surgery showed similar results. While the US, UK, and Germany are leading the efforts, research is concentrated in US institutes, including Johns Hopkins University, Harvard University, and Stanford University. The results for bioprinting showed that Chinese companies are getting ready to commercialize transformational

Table 5.1 Chapter highlights

		Leading countries	Leading institutions
Bibliometrics	COVID-19	US, China, Germany	University of Hong Kong, University of North Carolina, Harvard University
	Medical artificial intelligence	US, China, France	Harvard University, University of North Carolina, MIT
	Robotic surgery	US, UK, Germany	Johns Hopkins University, Harvard University, Stanford University
	Transgenic fish	US, Japan, Canada	Fisheries and Oceans Canada, Harvard University, Chinese Academy of Sciences
		Leading assignees	
Patent analysis	Bioprinting	Shenzhen Jiahong Oral Medical, Beijing Unicom Science and Technology, Hunan Huaxiang Incremental MFG	
	Medical 3D scanning	General Electric, Siemens Medical Solutions USA, Siemens Corporate Research	
	Wireless power	Witricity Corporation, Qualcomm, Energous Corporation	
	Drones in agriculture	Morris P. Kesler, Qualcomm, Andre B. Kurs	
		Leading institutions	
Network analysis	Automated vehicles	Carnegie Mellon University, University of Waterloo, Nagoya University, University of Western Australia	
	Electric vehicles	Tsinghua University, Aalborg University	
	Smart homes	King Saud University	
	Space travel	NASA, University of California, Berkeley, Russian Academy of Sciences	
		Leading countries	Leading institutions
Integrated analysis	Digital twin	US, People's Republic of China, Singapore	Beihang University, Guangdong University of Technology, National University of Singapore
	Supercomputing	US, UK, Germany	University of Illinois, Harvard University, University of California, Berkeley
		Leading assignees	
	Digital twin	General Electric, Siemens, Desktop Metal	
	Supercomputing	IBM, Intel, Rovi Guides	

technologies. Similarly, the results for medical 3D scanning suggested that companies who led prior technologies in the field are the ones investing in this technology. Those companies include General Electric and Siemens.

In the case of food and nutrition, intelligence gathered on transgenic fish showed that major institutes and countries are already engaged in this research. These countries include the US, Canada, China, and Japan. One common ocean all these countries share is the Pacific. We would need further data to determine whether this commonality has anything to do with our results. The data also suggest that drones are ready for commercialization in agriculture.

Transformation in transportation may have unprecedented impacts on our society. The intelligence gathered suggested that efforts are underway in almost every part of the developed world to transform urban mobility in terms of electric and automated vehicles. However, space travel is still lead by the US and Russia, who started the race to the moon decades ago. The data also suggested that other countries are coming up to speed in this area.

Smart homes and wireless power will transform how we live. We see efforts from different players in this case. We identified the presence of start-ups in the case of wireless power and a transforming economy in the case of smart homes. These cases show that it is critical to scan the start-up community as well as countries transforming from an old type of economy, like oil, to a future one, like smart homes, as in the case of Saudi Arabia.

There is more to be discussed in each case. Here, we have just provided high-level intelligence to demonstrate the power of the methods and applications included in this book.

References

Abraham, B. P., and Moitra, S. D. (2001). Innovation assessment through patent analysis. *Technovation*, 21(4), 245–252.

Akhan, S., and Canyurt, M. A. (2008). Transgenik Baliklar: Fayda ve Riskleri. *Journal of FisheriesSciences.com*, 2(3), 284–292.

Akiyama, M., Yamamoto, S., Fujita, K., Sakata, I., and Kajikawa, Y. (2012). Effective learning and knowledge discovery using processed medical incident reports. *Proceedings of PICMET '12 – Technology Management for Emerging Technologies*, 2337–2346.

Al, U., and Soydal, I. (2010). Bilgibilim alanında kendine atıf üzerine bir çalışma. *Bilgi dünyası*, 11(2), 349–364.

Al, U., Soydal, I., and Yalçın, H. (2010). An evaluation of the bibliometric features of Bilig. *Bilig*, 55, 1–20.

Alam, M. R., Reaz, M. B. I., and Ali, M. A. M. (2012). A review of smart homes – Past, present, and future. *IEEE Transactions on Systems, Man, and Cybernetics, Part C (Applications and Reviews)*, 42(6), 1190–1203.

Ameri, F., and Sabbagh, R. (2016). Digital factories for capability modeling and visualization. *Advances in Production Management Systems: Initiatives for a Sustainable World*, 488, 69–78.

An, J., Kim, K., Mortara, L., and Lee, S. (2018). Deriving technology intelligence from patents: Preposition-based semantic analysis. *Journal of Informetrics*, 12(1), 217–236.

Armanious, K., Abdulatif, S., Bhaktharaguttu, A. R., Küstner, T., Hepp, T., Gatidis, S., and Yang, B. (2021). Organ-based chronological age estimation based on 3D MRI scans. *Proceedings of the 28th European Signal Processing Conference (EUSIPCO)*, 1225–1228.

Bai, X., and Liu, Y. (2016). International collaboration patterns and effecting factors of emerging technologies. *PLoS One*, 11(12), e0167772.

Bakhtin, P., and Saritas, O. (2016). Tech mining for emerging STI trends through dynamic term clustering and semantic analysis: The case of photonics. In Daim, T. U., Chiavetta, D., Porter, A. L., Saritas, O. (eds), *Anticipating Future Innovation Pathways Through Large Data Analysis*. Berlin: Springer, 341–360.

Balta-Ozkan, N., Davidson, R., Bicket, M., and Whitmarsh, L. (2013). Social barriers to the adoption of smart homes. *Energy Policy*, 63, 363–374.

Bates, S. (2020). Literature listing. *World Patent Information*, 61, 101963.

Behkami, N., and Daim, T. U. (2012). Research forecasting for health information technology (HIT), using technology intelligence. *Technological Forecasting and Social Change*, 79(3), 498–508.

Berker, Y. (2018). Golden-ratio as a substitute to geometric and harmonic counting to determine multi-author publication credit. *Scientometrics*, 114(3), 839–857.

Biglu, M. H. (2009). Patent literature trends in Medline throughout 1965–2005. *Acimed*, 20(2).

Blue, R. S., Jennings, R. T., Antunano, M. J., and Mathers, C. H. (2017). Commercial spaceflight: Progress and challenges in expanding human access to space. *REACH*, 7–8, 6–13.
Blumrich, M. A., Chen, D., Chiu, G. L., Cipolla, T. M., Coteus, P. W., Gara, A. G., . . . Mok, L. S. (2009). *Massively Parallel Supercomputer*.
Blumrich, M. A., Chen, D., Chiu, G. L., Cipolla, T. M., Coteus, P. W., Gara, A. G., . . . Heidelberger, P. (2010). *Ultrascalable Petaflop Parallel Supercomputer*.
Bo, Q., Lichao, Z., Yusheng, S., and Guocheng, L. (2011). Support fast generation algorithm based on discrete-marking in stereolithgraphy rapid prototyping. *Rapid Prototyping Journal*, 17(6), 451–457.
Borbély, É. (2011). Could the SNA complete the SCOT model? Computer development in the USA between 1931–1950: A case study approach. *Periodica Polytechnica Social and Management Sciences*, 19(1), 25–36.
Boyack, K. W., Grafe, V. G., Johnson, D. K., and Wylie, B. N. (2002). *Patent Data Mining Method and Apparatus*.
Brandes, U., Delling, D., Gaertler, M., Gorke, R., Hoefer, M., Nikoloski, Z., and Wagner, D. (2007). On modularity clustering. *IEEE Transactions on Knowledge and Data Engineering*, 20(2), 172–188.
Brown, W. C. (1996). The history of wireless power transmission. *Solar Energy*, 56(1), 3–21.
Brundtland, G. H. (1987). Our common future – Call for action. *Environmental Conservation*, 14(4), 291–294.
Burmaoglu, S., Trajkovik, V., Tutukalo, T. L., Yalçın, H., and Caulfield, B. (2018). Evolution map of wearable technology patents for healthcare field. In Tong, R. (ed), *Wearable Technology in Medicine and Health Care*. Elsevier Science, 275–290.
Burnet, F. M., and White, D. O. (1972). *Natural History of Infectious Disease* (fourth edition). Cambridge: Cambridge University Press.
Burt, R. S. (1995). Social capital, structural holes and the entrepreneur. *Revue Francaise de Sociologie*, 36(4), 599– 628.
Cai, H., Xu, B., Jiang, L., and Vasilakos, A. V. (2016). IoT-based big data storage systems in cloud computing: Perspectives and challenges. *IEEE Internet of Things Journal*, 4(1), 75–87.
Campello, R. J., and Hruschka, E. R. (2006). A fuzzy extension of the silhouette width criterion for cluster analysis. *Fuzzy Sets and Systems*, 157(21), 2858–2875.
Canedo, A. (2016). Industrial IoT Lifecycle via Digital Twins. *2016 International Conference on Hardware/Software Codesign and System Synthesis (Codes+Isss)*.
Carrington, P. J., Scott, J., and Wasserman, S. (2005). *Models and Methods in Social Network Analysis*. Cambridge and New York: Cambridge University Press.
Carter-Templeton, H., Frazier, R. M., Wu, L., and Wyatt, T. (2018). Robotics in nursing: A bibliometric analysis. *Journal of Nursing Scholarship*, 50(6), 582–589.
Castelló-Cogollos, L., Sixto-Costoya, A., Lucas-Domínguez, R., Agulló-Calatayud, V., de Dios, J. G., and Aleixandre-Benavent, R. (2018). Bibliometría e indicadores de actividad científica (XI). Otros recursos útiles en la evaluación: Google Scholar, Microsoft Academic, 1findr, Dimensions y Lens.org. *Acta Pediatrica Espanola*, 76(9/10), 123–130.
Cavalheiro, M. B., Cavalheiro, G. M. D. C., Mayer, V. F., and Marques, O. R. B. (2021). Applying patent analytics to understand technological trends of smart tourism destinations. *Technology Analysis and Strategic Management*, 1–17.

Chang, S.-B. (2012). Using patent analysis to establish technological position: Two different strategic approaches. *Technological Forecasting and Social Change*, 79(1), 3–15.
Chen, C., Ibekwe-SanJuan, F., and Hou, J. (2010). The structure and dynamics of co-citation clusters: A multiple-perspective co-citation analysis. *Journal of the American Society for Information Science and Technology*, 61(7), 1386–1409.
Chen, T. T., and Powers, D. A. (1990). Transgenic fish. *Trends in Biotechnology*, 8, 209–215.
Chen, X., Chen, J., Wu, D., Xie, Y., and Li, J. (2016). Mapping the research trends by co-word analysis based on keywords from funded project. *Procedia Computer Science*, 91, 547–555.
Chen, X. L., Liu, Z. Q., Wei, L., Yan, J., Hao, T. Y., and Ding, R. Y. (2018). A comparative quantitative study of utilizing artificial intelligence on electronic health records in the USA and China during 2008–2017. *BMC Medical Informatics and Decision Making*, 18, 117.
Chen, Z., Luciani, A., Mateos, J. M., Barmettler, G., Giles, R. H., Neuhauss, S. C. F., and Devuyst, O. (2020). Transgenic zebrafish modeling low-molecular-weight proteinuria and lysosomal storage diseases. *Kidney International*, 97(6), 1150–1163.
Cho, R. L. T., Liu, J. S., and Ho, M. H. C. (2021). The development of autonomous driving technology: Perspectives from patent citation analysis. *Transport Reviews*, 1–27.
Cho, Y., and Daim, T. U. (2016). OLED TV technology forecasting using technology mining and the Fisher-Pry diffusion model. *Foresight*, 18(2), 117–137.
Choi, J. Y., Jeong, S., and Kim, K. (2015). A study on diffusion pattern of technology convergence: Patent analysis for Korea. *Sustainability*, 7(9), 11546–11569.
Choudhury, D., Anand, S., and Naing, M. W. (2018). The arrival of commercial bioprinters – Towards 3D bioprinting revolution. *International Journal of Bioprinting*, 4(2), 139.
Copeland, B. J. (2019). Artificial intelligence. *Encyclopædia Britannica*. Retrieved from https://www.britannica.com/technology/artificial-intelligence.
Coronado, D., Acosta, M., and Leon, D. (2004). Regional planning of R&D and science – technology interactions in Andalucia: A bibliometric analysis of patent documents. *European Planning Studies*, 12(8), 1075–1095.
Cowan, K. R., Daim, T. U., Wakeland, W., Fallah, H., Sheble, G., Lutzenhiser, L., . . . Nguyen, M. (2009). Forecasting the Adoption of Emerging Energy Technologies: Managing Climate Change and Evolving Social Values. *Proceedings of PICMET '09 – Technology Management in the Age of Fundamental Change*, 1–5, 2964.
Cox, A. M., Kennan, M. A., Lyon, L., Pinfield, S., and Sbaffi, L. (2019). Maturing research data services and the transformation of academic libraries. *Journal of Documentation*, 75(6), 1432–1462.
Crona, B., Ernstson, H., Prell, C., Reed, M., and Hubacek, K. (2011). Combining social network approaches with social theories to improve understanding of natural resource governance. In Bodin, O., and Prell, C. (eds), *Social Networks and Natural Resource Management: Uncovering the Social Fabric of Environmental Governance*. Cambridge: Cambridge University Press, 44–72.
Daim, T. U. (1997a). New product development (NPD) in high tech environments: A case study. *Proceedings of PICMET '97 – Innovation in Technology Management: The Key to Global Leadership*, 479–479.

Daim, T. U. (1997b). A review of evaluation of attributes for selecting advanced manufacturing technologies. *Proceedings of PICMET '97 – Innovation in Technology Management: The Key to Global Leadership*, 195–198.
Daim, T. U., Anderson, T., and Kocaoglu, D. (2015). Technology analytics: Enhancing technology assessment with technology intelligence. *Technological Forecasting and Social Change*, 97, 127–127.
Daim, T. U., Anderson, T. R., Thirumalai, M., Subramanian, G., Katarya, N., Krishnaswamy, D., and Singh, N. (2012). Data Center Technology Roadmap. In Luo, Z. (ed.), *Advanced Analytics for Green and Sustainable Economic Development: Supply Chain Models and Financial Technologies*. Pennsylvania: IGI Global, 202–230.
Daim, T. U., Chiavetta, D., Porter, A. L., and Saritas, O. (eds) (2016). *Anticipating Future Innovation Pathways through Large Data Analysis*. Berlin: Springer.
Daim, T. U., Grueda, G., Martin, H., and Gerdsri, P. (2006). Forecasting emerging technologies: Use of bibliometrics and patent analysis. *Technological Forecasting and Social Change*, 73(8), 981–1012.
Daim, T. U., Iskin, I., Li, X., Zielsdorff, C., Bayraktaroglu, A. E., Dereli, T., and Durmusoglu, A. (2012). Patent analysis of wind energy technology using the patent alert system. *World Patent Information*, 34(1), 37–47.
Daim, T. U., Lai, K. K., Yalçın, H., Alsoubie, F., and Kumar, V. (2020). Forecasting technological positioning through technology knowledge redundancy: Patent citation analysis of IoT, cybersecurity, and blockchain. *Technological Forecasting and Social Change*, 161, 120329.
Daim, T. U., Monalisa, M., Pranabesh, D., and Brown, N. (2007). Time lag assessment between research funding and output in emerging technologies: Case of scope applications. *Foresight*, 9(4), 33–44.
Daim, T. U., et al. (2012). Patent analysis of wind energy technology using the patent alert system. *World Patent Information*, 34(1), 37–47.
Daim, T. U., and Suntharasaj, P. (2009). Technology diffusion: Forecasting with bibliometric analysis and Bass model. *Foresight*, 11(3), 45–55.
Daim, T. U., and Yalçın, H. (2019). The main sources for technology management research: A bibliometric approach. In Kocaoglu, D. F., Anderson, T. R., Kozanoglu, D. C., Niwa, K., and Steenhuis, H. J. (eds), *2019 Portland International Conference on Management of Engineering and Technology*. New York: IEEE.
de Araújo Boleti, A. P., de Oliveira Flores, T. M., Moreno, S. E., Dos Anjos, L., Mortari, M. R., and Migliolo, L. (2020). Neuroinflammation: An overview of neurodegenerative and metabolic diseases and of biotechnological studies. *Neurochemistry International*, 136, 104714.
De Bruin, R., and Moed, H. (1993). Delimitation of scientific subfields using cognitive words from corporate addresses in scientific publications. *Scientometrics*, 26(1), 65–80.
De Silva, L. C., Morikawa, C., and Petra, I. M. (2012). State of the art of smart homes. *Engineering Applications of Artificial Intelligence*, 25(7), 1313–1321.
de Stefano, E., de Sequeira Santos, M. P., and Balassiano, R. (2016). Development of a software for metric studies of transportation engineering journals. *Scientometrics*, 109(3), 1579–1591.
DebRoy, T., Zhang, W., Turner, J., and Babu, S. S. (2017). Building digital twins of 3D printing machines. *Scripta Materialia*, 135, 119–124.
Delbrugger, T., Lenz, L. T., Losch, D., and Rossmann, J. (2017). A navigation framework for digital twins of factories based on building information modeling. *22nd*

IEEE International Conference on Emerging Technologies and Factory Automation (Etfa).

Den, Y., Koiso, H., Maruyama, T., Maekawa, K., Takanashi, K., Enomoto, M., and Yoshida, N. (2010). Two-level annotation of utterance-units in Japanese dialogs: An empirically emerged scheme. Paper presented at the LREC.

Ding, Y., Chowdhury, G. G., and Foo, S. (2001). Bibliometric cartography of information retrieval research by using co-word analysis. *Information Processing and Management*, 37(6), 817–842.

do Prado, J. W., de Castro Alcântara, V., de Melo Carvalho, F., Vieira, K. C., Machado, L. K. C., and Tonelli, D. F. (2016). Multivariate analysis of credit risk and bankruptcy research data: A bibliometric study involving different knowledge fields (1968–2014). *Scientometrics*, 106(3), 1007–1029.

Duan, C. H. (2011). Mapping the intellectual structure of modern technology management. *Technology Analysis and Strategic Management*, 23(5), 583–600.

Dunham, R. A. (2011). *Aquaculture and Fisheries Biotechnology: Genetic Approaches*. Wallingford: CABI.

Ebrahim, T. Y. (2017). 3D Bioprinting Patentable Subject Matter Boundaries. *Seattle University Law Review*, 41(1).

Egghe, L., and Rousseau, R. (2006). An informetric model for the Hirsch-index. *Scientometrics*, 69(1), 121–129.

Ehrlich, K., and Carboni, I. (2005). Inside social network analysis. IBM technical report.

Ekici, A., Timur, M., and Bağış, H. (2006). Transgenik Canlılar ve Akuakültürdeki Önemi. *Su Ürünleri Dergisi*, 23(2), 211–214.

Eom, S. B. (1995). Decision-support systems research: Reference disciplines and a cumulative tradition. *Omega – International Journal of Management Science*, 23(5), 511–523.

Eom, S. B. (1996). Mapping the intellectual structure of research in decision support systems through author co-citation analysis (1971–1993). *Decision Support Systems*, 16(4), 315–338.

Ewing, G. J. (1966). Citation of articles from volume 58 of the Journal of Physical Chemistry. *Journal of Chemical Documentation*, 6(4), 247–250.

Fallah, H., Choudhury, P., and Daim, T. U. (2012). Does movement of inventors between companies affect their productivity? *Technology in Society*, 34(3), 196–206.

Fan, G., Zhou, Z., Zhang, H., Gu, X., Gu, G., Guan, X., ... and He, S. (2016). Global scientific production of robotic surgery in medicine: A 20-year survey of research activities. *International Journal of Surgery*, 30, 126–131.

Feng, J., Liu, Z., and Feng, L. (2021). Identifying opportunities for sustainable business models in manufacturing: Application of patent analysis and generative topographic mapping. *Sustainable Production and Consumption*, 27, 509–522.

Fiala, D., and Willett, P. (2015). Computer science in Eastern Europe 1989–2014: A bibliometric study. *Aslib Journal of Information Management*, 67(5), 526–541.

Freedonia Group (2006). *Hybrid–Electric Vehicles and Competing Automotive Powerplants*. Cleveland: Freedonia Group.

Freedonia Group (2014). *World Hybrid and Electric Vehicles*. Cleveland: Freedonia Group.

Gapinski, B., Wieczorowski, M., Marciniak-Podsadna, L., Dybala, B., and Ziolkowski, G. (2014). Comparison of different method of measurement geometry using CMM, optical scanner and computed tomography 3D. *Procedia Engineering*, 69, 255–262.

Garces, E., van Blommestein, K., Anthony, J., Hillegas-Elting, J., Daim, T. U., and Yoon, B. (2017). Technology domain analysis: A case of energy-efficient advanced commercial refrigeration technologies. *Sustainable Production and Consumption*, 12, 221–233.

Garechana, G., Rio-Belver, R., Bildosola, I., and Cilleruelo-Carrasco, E. (2019). A method for the detection and characterization of technology fronts: Analysis of the dynamics of technological change in 3D printing technology. *PLoS One*, 14(1), e0210441.

Garfield, E. (1979). Is citation analysis a legitimate evaluation tool? *Scientometrics*, 1(4), 359–375.

Garfield, E. (2006). The history and meaning of the journal impact factor. *JAMA*, 295(1), 90–93.

Garfield, E., Sher, I. H., and Torpie, R. J. (1964). *The Use of Citation Data in Writing the History of Science*. Philadelphia: Institute for Scientific Information.

Gibson, E., Blommestein, K., Kim, J., Daim, T. U., and Garces, E. (2017). Forecasting the electric transformation in transportation. *Technology Analysis and Strategic Management*, 29(10), 1103–1120.

Glaessgen, E., and Stargel, D. (2012). The digital twin paradigm for future NASA and US Air Force vehicles. *53rd AIAA/ASME/ASCE/AHS/ASC Structures, Structural Dynamics, and Materials Conference*, Honolulu, Hawaii.

Glänzel, W., and Schubert, A. (2003). A new classification scheme of science fields and subfields designed for scientometric evaluation purposes. *Scientometrics*, 56(3), 357–367.

Glass, I. I. (1974). Aerospace in the next century. *Progress in Aerospace Sciences*, 15, 257–324.

Gonçalves Pereira, C., Lavoie, J., Garces, E., Basso, F., Dabić, M., Porto, G., and Daim, T. U. (2019). Assessment of technologies: forecasting of emerging therapeutic monoclonal antibodies patents based on a decision model. *Technological Forecasting and Social Change*, 139, 185–199.

Gopinathan, J., and Noh, I. (2018). Intellectual properties in applications of biomimetic medical materials: Current status of development and intellectual properties of biomimetic medical materials. In Noh, I. (ed.), *Biomimetic Medical Materials*. Berlin: Springer, 377–399.

Gourin, C. G., and Terris, D. J. (2007). *History of Robotic Surgery. Robotics in Surgery: History, Current and Future applications*. New York: Nova Science Publishers, 3–12.

Guan, J., Zuo, K., Chen, K., and Yam, R. C. (2016). Does country-level R&D efficiency benefit from the collaboration network structure? *Research Policy*, 45(4), 770–784.

Gupta, B. M., and Dhawan, S. M. (2018). Artificial intelligence research in India: A scientometric assessment of publications output during 2007–16. *Desidoc Journal of Library and Information Technology*, 38(6), 416–422.

Haleem, A., and Javaid, M. (2019). 3D scanning applications in medical field: A literature-based review. *Clinical Epidemiology and Global Health*, 7(2), 199–210.

Han, X. S., Li, Y., Xue, X. L., and Shi, N. (2012). *A Framework for Construction Innovation Research Based on Network Analysis*. Beijing: China Architecture and Building Press.

Harnad, S. (2006). The annotation game: On Turing (1950) on computing, machinery, and intelligence. In Epstein, R., and Peters, G. (eds), *Parsing the Turing Test: Philosophical and Methodological Issues in the Quest for the Thinking Computer*: Kluwer.

Heinzerling, T., Oppelt, M., and Bell, T. (2017). Integration of device models into the process simulation – Concept and evaluation of a model description. *Atp Edition*, 10, 34–45.

Hinze, S., and Grupp, H. (1996). Mapping of R&D structures in transdisciplinary areas. New biotechnology in food sciences. *Scientometrics*, 37(2), 313–335.

Hu, J., and Zhang, Y. (2015). Research patterns and trends of recommendation system in China using co-word analysis. *Information Processing and Management*, 51(4), 329–339.

Hua, K., and Ekker, M. (2020). Life, death, and regeneration of zebrafish dopaminergic neurons. In Gerlai, R. T. (ed.), *Behavioral and Neural Genetics of Zebrafish*. Cambridge: Academic Press, 363–376.

Huang, Z., Chen, H., Yip, A., Ng, G., Guo, F., Chen, Z.-K., and Roco, M. C. (2003). Longitudinal patent analysis for nanoscale science and engineering: Country, institution and technology field. *Journal of Nanoparticle Research*, 5(3–4), 333–363.

Hueso, M., Vellido, A., Montero, N., Barbieri, C., Ramos, R., Angoso, M., . . . Jonsson, A. (2018). Artificial intelligence for the artificial kidney: Pointers to the future of a personalized hemodialysis therapy. *Kidney Diseases*, 4(1), 1–9.

Hüls, J., and Remke, A. (2017). Coordinated charging strategies for plug-in electric vehicles to ensure a robust charging process. *10th EAI International Conference on Performance Evaluation Methodologies and Tools*.

IBM (2019). Watson Health: Get the facts. Retrieved from https://www.ibm.com/watson-health/about/get-the-facts.

International Energy Agency (2007a). *Implementing Agreement for Hybrid and Electric Vehicle Technologies and Programmes*. Paris: International Energy Agency.

International Energy Agency (2007b). *Hybrid and Electric Vehicles: Past, Present, Future*. Paris: International Energy Agency.

Jackson, S. R., and Patel, M. I. (2019). Robotic surgery research in urology: A bibliometric analysis of field and top 100 articles. *Journal of Endourology*, 33(5), 389–395.

Jacsó, P. (2009). The h-index for countries in Web of Science and Scopus. *Online Information Review*, 33(4), 831–837.

Jain, N., and Srivastava, V. (2013). Data mining techniques: a survey paper. *International Journal of Research in Engineering and Technology*, 2(11), 2319–1163.

Jeong, C., and Kim, K. (2014). Creating patents on the new technology using analogy-based patent mining. *Expert Systems with Applications*, 41(8), 3605–3614.

Jun, S., Han, S. H., Yu, J., Hwang, J., Kim, S., and Lee, C. (2021). Identification of promising vacant technologies for the development of truck on freight train transportation systems. *Applied Sciences*, 11(2), 499.

Jun, S., and Park, S. S. (2013). Examining technological innovation of Apple using patent analysis. *Industrial Management and Data Systems*, 113(6), 890–907.

Jung, S. Y., Lee, S. J., Kim, H. Y., Park, H. S., Wang, Z., Kim, H. J., . . . Kim, H. S. (2016). 3D printed polyurethane prosthesis for partial tracheal reconstruction: A pilot animal study. *Biofabrication*, 8(4), 045015.

Kadry, S., and Al-Taie, M. Z. (2014). *Social Network Analysis: An Introduction with an Extensive Implementation to a Large-Scale Online Network Using Pajek*. Sharjah: Bentham Science Publishers.

Kantardzic, M. (2011). *Data Mining: Concepts, Models, Methods, and Algorithms*. New York: John Wiley and Sons.

Karvonen, M., and Kässi, T. (2011). Patent analysis for analysing technological convergence. *Foresight*, 13(5), 34–50.

Kellner, M., Heidrich, M., Lorbeer, R. A., Antonopoulos, G. C., Knudsen, L., Wrede, C., . . . Meyer, H. (2016). A combined method for correlative 3D imaging of biological samples from macro to nano scale. *Scientific Reports*, 6, 35606.

Khalil, S., El-Badri, N., El-Mokhtaar, M., Al-Mofty, S., Farghaly, M., Ayman, R., . . . Mousa, N. (2016). A cost-effective method to assemble biomimetic 3D cell culture platforms. *PLoS One*, 11(12), e0167116.

Khan, G. F., and Wood, J. (2015). Information technology management domain: Emerging themes and keyword analysis. *Scientometrics*, 105(2), 959–972.

Koito, H., Suzuki, J., Ohkubo, N., Ishiguro, Y., Iwasaka, T., Inada, M., and Nakano, Y. (1996). Three-dimensional reconstructed magnetic resonance imaging for diagnosing persistent left superior vena cava: Comparison with magnetic resonance angiography and plain chest radiography. *Journal of Cardiology*, 28(3), 161–170.

Köpping Athanasopoulos, H. (2019). The Moon Village and Space 4.0: The 'Open Concept' as a new way of doing space? *Space Policy*, 49, 101323.

Koren, G., Polack, F., and Joyeux, D. (1993). Digital twin image elimination in soft-X-Ray in-line holography. *Soft X-Ray Microscopy*, 1741, 260–274.

Kubina, M., Varmus, M., and Kubinova, I. (2015). Use of big data for competitive advantage of company. *Procedia Economics and Finance*, 26, 561–565.

Kumar, S., and Kumar, S. (2008). Collaboration in research productivity in oil seed research institutes of India. *Fourth International Conference on Webometrics, Informetrics and Scientometrics*.

Lang, U., Peltier, J. P., Christ, P., Rill, S., Rantzau, D., Nebel, H., . . . Haas, P. (1995). Perspectives of collaborative supercomputing and networking in European Aerospace research and industry. *Future Generation Computer Systems*, 11(4), 419–430.

Larina, I. M., Pastushkova, L. K., Kononikhin, A. S., Nikolaev, E. N., and Orlov, O. I. (2019). Piloted space flight and post-genomic technologies. *REACH*, 16, 100034.

Launius, R. D. (2008). Space stations for the United States: An idea whose time has come – and gone? *Acta Astronautica*, 62(10), 539–555.

Le, M. Q., Capsal, J.-F., Lallart, M., Hebrard, Y., Van Der Ham, A., Reffe, N., . . . Cottinet, P.-J. (2015). Review on energy harvesting for structural health monitoring in aeronautical applications. *Progress in Aerospace Sciences*, 79, 147–157.

Lee, P. C., and Su, H. N. (2011). Quantitative mapping of scientific research–The case of electrical conducting polymer nanocomposite. *Technological Forecasting and Social Change*, 78(1), 132–151.

Lewison, G. (1996). The definition of biomedical research subfields with title keywords and application to the analysis of research outputs. *Research Evaluation*, 6(1), 25–36.

Li, S., Garces, E., and Daim, T. U. (2019). Technology forecasting by analogy based on social network analysis: The case of autonomous vehicles. *Technological Forecasting and Social Change*, 148, 119731.

Li, Y.-R., Wang, L.-H., and Hong, C.-F. (2009). Extracting the significant-rare keywords for patent analysis. *Expert Systems with Applications*, 36(3), 5200–5204.

Liang, Y. X., and Li, Z. F. (2010). Bibliometrics analysis of science and technology policy in China. *Proceedings of the 7th International Conference on Innovation and Management*, I and II, 1339–1343.

Lin, X., Xie, Q., Daim, T. U., and Huang, L. (2019). Forecasting technology trends using text mining of the gaps between science and technology: The case of perovskite solar cell technology. *Technological Forecasting and Social Change*, 146, 432–449.

Liu, Z., Moav, B., Faras, A. J., Guise, K. S., Kapuscinski, A. R., and Hackett, P. B. (1990). Development of expression vectors for transgenic fish. *Bio/Technology*, 8(12), 1268–1272.

Lurie, N., Saville, M., Hatchett, R., and Halton, J. (2020). Developing Covid-19 vaccines at pandemic speed. *New England Journal of Medicine*, 382, 1969–1973.

Machado, C., and Davim, J. P. (2020). *Industry 4.0: Challenges, Trends, and Solutions in Management and Engineering*. Boca Raton: CRC Press/Taylor and Francis Group.

Madani, F. (2015). 'Technology mining' bibliometrics analysis: applying network analysis and cluster analysis. *Scientometrics*, 105(1), 323–335.

Madani, F., Daim, T. U., and Weng, C. (2017). Smart building technology network analysis: Applying core periphery structure analysis. *International Journal of Management Science and Engineering Management*, 12(1), 1–11.

Madani, F., Daim, T. U., and Zwick, M. (2018). Keyword-based patent citation prediction via information theory. *International Journal of General Systems*, 47(8), 821–841.

Madani, F., and Weber, C. (2016). The evolution of patent mining: Applying bibliometrics analysis and keyword network analysis. *World Patent Information*, 46, 32–48.

Maeno, T., Shibata, N., Kajikawa, Y., and Sakata, I. (2011). Investigation of a lead indicator of technological innovations. *Proceedings of PICMET '11: Technology Management in the Energy Smart World*, 1–6.

Manohar, M., Lathabai, H. H., George, S., and Prabhakaran, T. (2018). Wire-free electricity: Insights from a techno-futuristic exploration. *Utilities Policy*, 53, 3–14.

Marjani, M., Nasaruddin, F., Gani, A., Karim, A., Hashem, I. A. T., Siddiqa, A., and Yaqoob, I. (2017). Big IoT data analytics: architecture, opportunities, and open research challenges. *IEEE Access*, 5, 5247–5261.

Martin, H., and Daim, T. U. (2012). Technology roadmap development process (TRDP) for the service sector: A conceptual framework. *Technology in Society*, 34(1), 94–105.

Marzi, G., Dabic, M., Daim, T. U., and Garces, E. (2017). Product and process innovation in manufacturing firms – A thirty-year bibliometric analysis. *Scientometrics*, 113(2), 673–704.

McCain, K. W. (1995). The structure of biotechnology R&D. *Scientometrics*, 32(2), 153–175.

McCarthy, J., Minsky, M., Rochester, N., and Shannon, C. (1955). A proposal for the Dartmouth summer conference on artificial intelligence. *AI Magazine*, 31.

McCarty, C., Jawitz, J. W., Hopkins, A., and Goldman, A. (2013). Predicting author h-index using characteristics of the co-author network. *Scientometrics*, 96(2), 467–483.

McCulloch, W. S., and Pitts, W. (1943). A logical calculus of the ideas immanent in nervous activity. *Bulletin of Mathematical Biophysics*, 5, 115–133. Reprinted in *Bulletin of Mathematical Biology*, 52(1–2), 99–115, 1990.

MedlinePlus (2020). *X-rays*. Retrieved from https://medlineplus.gov/xrays.html.

Meyer, M. S. (2000). Patent citations in a novel field of technology: What can they tell about interactions between emerging communities of science and technology? *Scientometrics*, 48(2), 151–178.

Meyer, M. S. (2001). Patent citation analysis in a novel field of technology: An exploration of nano-science and nano-technology. *Scientometrics*, 51(1), 163–183.

Minssen, T., and Mimler, M. (2016). Patenting bioprinting-technologies in the US and Europe – The 5th element in the 3rd dimension. Working paper.

Moed, H. F. (2005). Citation analysis of scientific journals and journal impact measures. *Current Science*, 89(12), 1990–1996.

Moed, H. F., Glänzel, W., and Schmoch, U. (2004). *Handbook of Quantitative Science and Technology Research*. Amsterdam: Kluwer Academic, 257–276.

Morimoto, A. K., Krumm, J. C., Kozlowski, D. M., Kuhlmann, J. L., Wilson, C., Little, C., . . . Walsh, N. (1997). High definition 3D ultrasound imaging. *Studies in Health Technology and Informatics*, 39, 90–98.

Mu-Hsuan, H., and Hsiao-Wen, Y. (2013). A scientometric study of fuel cell based on paper and patent analysis. *Journal of Library and Information Studies*, 11(1), 1–24.

Narin, F., Pinski, G., and Gee, H. H. (1976). Structure of the biomedical literature. *Journal of the American Society for Information Science*, 27(1), 25–45.

National Institute of Biomedical Imaging and Bioengineering (2020). *Computed tomography (CT)*. Retrieved from https://www.nibib.nih.gov/science-education/science-topics/computed-tomography-ct.

Naumanen, M., Uusitalo, T., Huttunen-Saarivirta, E., and van der Have, R. (2019). Development strategies for heavy duty electric battery vehicles: Comparison between China, EU, Japan and USA. *Resources, Conservation and Recycling*, 151, 104413.

Nooy, W. D., Mrvar, A., and Batagelj, V. (2011). *Exploratory Social Network Analysis with Pajek* (second edition). New York: Cambridge University Press.

Nooy, W. D., Mrvar, A., and Batagelj, V. (2018). *Exploratory Social Network Analysis with Pajek* (third edition). New York: Cambridge University Press.

Noyons, E., and Van Raan, A. (1994). Bibliometric cartography of scientific and technological developments of an R&D field: The case of optomechatronics. *Scientometrics*, 30(1), 157–173.

OECD (2005). Oslo manual: Guidelines for collecting and interpreting innovation data. *Measurement of Scientific and Technological Activities*.

Olvera, C., Berbegal-Mirabent, J., and Merigo, J. M. (2018). A bibliometric overview of university–business collaboration between 1980 and 2016. *Computacion Y Sistemas*, 22(4), 1171–1190.

Otte, E., and Rousseau, R. (2002). Social network analysis: A powerful strategy, also for the information sciences. *Journal of Information Science*, 28(6), 441–453.

Özden, O., Güner, Y., and Kızak, V. (2003). Tatlısu Balık Kültüründe Uygulanan Bazı Biyoteknolojik Yöntemler. *Su Ürünleri Dergisi*, 20(3).

Pantano, E., Priporas, C.-V., Sorace, S., and Iazzolino, G. (2017). Does innovation-orientation lead to retail industry growth? Empirical evidence from patent analysis. *Journal of Retailing and Consumer Services*, 34, 88–94.

Pao-Long, C., and Hoang-Jyh, L. (2010). Using patent analyses to monitor the technological trends in an emerging field of technology: A case of carbon nanotube field emission display. *Scientometrics*, 82(1), 5–19.

Park, H. W., and Thelwall, M. (2008a). Link analysis: Hyperlink patterns and social structure on politicians' web sites in South Korea. *Quality and Quantity*, 42(5), 687–697.

Park, H. W., and Thelwall, M. (2008b). Developing network indicators for ideological landscapes from the political blogosphere in South Korea. *Journal of Computer-Mediated Communication*, 13(4), 856–879.

Pelicioni, L. C., Ribeiro, J. R., Devezas, T., Belderrain, M. C. N., and Melo, F. C. L. D. (2018). Application of a bibliometric tool for studying space technology trends. *Journal of Aerospace Technology and Management*, 10, e0318.

Petrovic, S. (2006). A comparison between the silhouette index and the Davies–Bouldin index in labelling IDS clusters. *11th Nordic Workshop of Secure IT Systems*.
Pietrzyk, J., Habich, D., Damme, P., and Lehner, W. (2019). First investigations of the vector supercomputer SX-Aurora TSUBASA as a co-processor for database systems. *BTW 2019 – Workshopband*, Bonn, Germany.
Pilkington, A., and Teichert, T. (2006). Management of technology: Themes, concepts and relationships. *Technovation*, 26(3), 288–299.
Popova, N. K., Kulikov, A. V., and Naumenko, V. S. (2020). Spaceflight and brain plasticity: Spaceflight effects on regional expression of neurotransmitter systems and neurotrophic factors encoding genes. *Neuroscience and Biobehavioral Reviews*, 119, 396–405.
Porter, A. L., and Cunningham, S. W. (2004). *Tech Mining: Exploiting New Technologies for Competitive Advantage*. New York: John Wiley and Sons.
Pouris, A. (1991). Identifying areas of strength in South African technology. *Scientometrics*, 21(1), 23–35.
Prescott, M. E. (2014). Big data and competitive advantage at Nielsen. *Management Decision*, 52(3), 573–601.
Pritchard, A. (1969). Statistical bibliography or bibliometrics. *Journal of Documentation*, 25(4), 348–349.
Puig, J., and Duran, J. (2010). Digital twins. *4th International Multi-Conference on Society, Cybernetics and Informatics*, II, 28–31.
Rajeswari, A. R. (1996). Indian patent statistics – An analysis. *Scientometrics*, 36(1), 109–130.
Ramani, S. V., de Looze, M.-A., Davis, M., and Wilson, C. S. (2001). Using patent statistics as competition indicators in the biotechnology sectors: An application to France, Germany and the UK. *Proceedings of the 8th International Conference on Scientometrics and Informetrics*, 543–555.
Ramirez Barreto, D. A., Guillermo, O., Valencia, E., Peña Rodriguez, A., and Cardenas Escorcia, Y. D. C. (2018). Bibliometric analysis of nearly a decade of research in electric vehicles: A dynamic approach. *ARPN Journal of Engineering and Applied Sciences*, 12(17), 4730–4736.
Rand, D. A. J., Woods, R., and Dell, R. (1998). *Batteries for Electric Vehicles*. Somerset: Research Studies Press; New York: John Wiley and Sons.
Reinisch, C., Kofler, M., Iglesias, F., and Kastner, W. (2011). Thinkhome energy efficiency in future smart homes. *EURASIP Journal on Embedded Systems*, 2011(1), 104617.
Renaudin, C. P., Barbier, B., Roriz, R., Revel, D., and Amiel, M. (1994). Coronary arteries: New design for three-dimensional arterial phantoms. *Radiology*, 190(2), 579–582.
Rio-Belver, R., and Cilleruelo, E. (2010). Discovering technologies using techmining: The case of waste recycling. *6th International Scientific Conference Business and Management*, I and II, 950–955.
Rio-Belver, R., Garechana, G., Bildosola, I., and Zarrabeitia, E. (2018). Evolution and scientific visualization of machine learning field. *Proceedings of the 2nd International Conference on Advanced Research Methods and Analytics*, 115–123.
Rivas, J. M., Gutiérrez, J. J., and González Harbour, M. (2017). A supercomputing framework for the evaluation of real-time analysis and optimization techniques. *Journal of Systems and Software*, 124, 120–136.
Rodic, B. (2017). Industry 4.0 and the new simulation modelling paradigm. *Organizacija*, 50(3), 193–207.

Rodriguez-Salvador, M., Rio-Belver, R. M., and Garechana-Anacabe, G. (2017). Scientometric and patentometric analyses to determine the knowledge landscape in innovative technologies: The case of 3D bioprinting. *PLoS One*, 12(6), e0180375.

Rosenberg, C. E. (1989). What is an epidemic? AIDS in historical perspective. *Daedalus*, 118(2), 1–17.

Russom, P. (2011). Big data analytics. Report, TDWI.

Sabidussi, G. (1966). The centrality index of a graph. *Psychometrika*, 31(4), 581–603.

Sannita, W. G., Narici, L., and Picozza, P. (2006). Positive visual phenomena in space: A scientific case and a safety issue in space travel. *Vision Research*, 46(14), 2159–2165.

Saputra, L. K. P., and Lukito, Y. (2017). Implementation of air conditioning control system using REST protocol based on NodeMCU ESP8266. *International Conference on Smart Cities, Automation and Intelligent Computing Systems (ICON-SONICS),* Yogyakarta, Indonesia.

Saritas, O., and Burmaoglu, S. (2016). Future of sustainable military operations under emerging energy and security considerations. *Technological Forecasting and Social Change*, 102, 331–343.

Schilling, M. A., and Shankar, R. (2019). *Strategic Management of Technological Innovation*. New York: McGraw-Hill Education.

Schröder, V., and Widera, P. (2021). Obtaining evidence in patent litigation and trade secret protection – A tale of two poles. *GRUR International*, 70(4), 361–376.

Schroeder, G. N., Steinmetz, C., Pereira, C. E., and Espindola, D. B. (2016). Digital twin data modeling with AutomationML and a communication methodology for data exchange. *IFAC-PapersOnLine*, 49(30), 12–17.

Schuh, G., and Blum, M. (2016). Design of a data structure for the order processing as a basis for data analytics methods. *Proceedings of PICMET '16 – Technology Management for Social Innovation*, 2164–2169.

Scott, J. (2000). *Social Network Analysis: A Handbook* (second edition). London and Thousand Oaks: SAGE Publications.

Schwartz, S., and Hellin, J. (1996). Measuring the impact of scientific publications. The case of the biomedical sciences. *Scientometrics*, 35(1), 119–132.

Shen, L., Wang, S., Dai, W., and Zhang, Z. (2019). Detecting the interdisciplinary nature and topic hotspots of robotics in surgery: Social network analysis and bibliometric study. *Journal of Medical Internet Research*, 21(3), e12625.

Shibata, N., Kajikawa, Y., and Sakata, I. (2011). Detecting potential technological fronts by comparing scientific papers and patents. *Foresight*, 13(5), 51–60.

Simmons, A., Fellous, J.-L., Ramaswamy, V., Trenberth, K., Asrar, G., Balmaseda, M., . . . Shepherd, T. (2016). Observation and integrated Earth-system science: A roadmap for 2016–2025. *Advances in Space Research*, 57(10), 2037–2103.

Still, K., Huhtamaki, J., Russell, M. G., and Rubens, N. (2014). Insights for orchestrating innovation ecosystems: The case of EIT ICT Labs and data-driven network visualisations. *International Journal of Technology Management*, 66(2–3), 243–265.

Sumi, F. H., Dutta, L., and Sarker, F. (2018). Future with wireless power transfer technology. *Journal of Electrical and Electronic Systems*, 7(279).

Sun, H., Geng, Y., Hu, L., Shi, L., and Xu, T. (2018). Measuring China's new energy vehicle patents: A social network analysis approach. *Energy*, 153, 685–693.

Sydor, M. (2019). Geometry of wood screws: A patent review. *European Journal of Wood and Wood Products*, 77(1), 93–103.

Szocik, K., Wójtowicz, T., and Baran, L. (2017). War or peace? The possible scenarios of colonising Mars. *Space Policy*, 42, 31–36.

Tachi, K., Furukawa, K. S., Koshima, I., and Ushida, T. (2011). New microvascular anastomotic ring–coupling device using negative pressure. *Journal of Plastic, Reconstructive and Aesthetic Surgery*, 64(9), 1187–1193.

Tian, X., Geng, Y., Zhong, S., Wilson, J., Gao, C., Chen, W., . . . Hao, H. (2018). A bibliometric analysis on trends and characters of carbon emissions from transport sector. *Transportation Research Part D: Transport and Environment*, 59, 1–10.

Tomasev, N., Glorot, X., Rae, J. W., Zielinski, M., Askham, H., Saraiva, A., . . . Mohamed, S. (2019). A clinically applicable approach to continuous prediction of future acute kidney injury. *Nature*, 572(7767), 116–119.

Topalli, M., and Ivanaj, S. (2016). Mapping the evolution of the impact of economic transition on Central and Eastern European enterprises: A co-word analysis. *Journal of World Business*, 51(5), 744–759.

Tseng, Y.-H., Lin, C.-J., and Lin, Y.-I. (2007). Text mining techniques for patent analysis. *Information Processing and Management*, 43(5), 1216–1247.

Umut, A. L., and Coştur, R. (2007). Türk Psikoloji Dergisi'nin bibliyometrik profili. *Türk kütüphaneciliği*, 21(2), 142–163.

Ustundag, M. T., Yalçın, H., and Gunes, E. (2016). Intellectual structure of stem education in educational research. *Turkish Online Journal of Educational Technology*, November special edition, 1222–1230.

Van Eck, N., and Waltman, L. J. S. (2009). Software survey: VOSviewer, a computer program for bibliometric mapping. *Scientometrics*, 84(2), 523–538.

Verbeek, A., Debackere, K., Luwel, M., Andries, P., Davis, M., and Wilson, C. S. (2001). Linking science to technology: using bibliographic references in patents to build linkage schemes. *Proceedings of the 8th International Conference on Scientometrics and Informetrics*, 717–732.

Vicente-Gomila, J. M., Palli, A., de la Calle, B., Artacho, M. A., and Jimenez, S. (2017). Discovering shifts in competitive strategies in probiotics, accelerated with TechMining. *Scientometrics*, 111(3), 1907–1923.

Vickery, B. C. (1948). Bradford's law of scattering. *Journal of Documentation*, 4(3), 198–203.

Waltman, L., and Van Eck, N. J. (2012). A new methodology for constructing a publication-level classification system of science. *Journal of the American Society for Information Science and Technology*, 63(12), 2378–2392.

Wang, K. J., Lee, T. L., and Hsu, Y. L. (2020). Revolution on digital twin technology – A patent research approach. *International Journal of Advanced Manufacturing Technology*, 107(11–12), 4687–4704.

Wasserman, S., and Faust, K. (1994). *Social Network Analysis: Methods and Applications*. Cambridge and New York: Cambridge University Press.

Wasserman, S., and Galaskiewicz, J. (1994). *Advances in Social Network Analysis: Research in the Social and Behavioral Sciences*. Thousand Oaks: Sage Publications.

Weiser, M., and Brown, J. S. (1996). Designing calm technology. *PowerGrid Journal*, 1(1), 75–85.

Weiser, M., and Brown, J. S. (1997). The coming age of calm technology. In Denning, P., and Metcalfe, R. (eds), *Beyond Calculation*. Berlin: Springer, 75–85.

Weng, C. S. (2014). Technology management: The perspective of social network. *International Journal of Innovation and Technology Management*, 11(3), 7.

Wikipedia (2019). *Digital twin*. Retrieved from https://en.wikipedia.org/wiki/Digital_twin.

Wikipedia (2020a). *Magnetic resonance imaging*. Retrieved from https://en.wikipedia.org/wiki/Magnetic_resonance_imaging.

Wikipedia (2020b). *Medical ultrasound*. Retrieved from https://en.wikipedia.org/wiki/Medical_ultrasound.

Wikipedia (2021a). *Argonne National Laboratory*. Retrieved from https://en.wikipedia.org/wiki/Argonne_National_Laboratory.

Wikipedia (2021b). *Oak Ridge National Laboratory*. Retrieved from https://en.wikipedia.org/wiki/Oak_Ridge_National_Laboratory.

Wilson, R. M. (1987). Patent analysis using online databases – I. Technological trend analysis. *World Patent Information*, 9(1), 18–26.

Witten, M. (1991). Supercomputers in biology and medicine – An overview. *Supercomputer*, 8(2), 37–53.

Wittfoth, S. (2019). Measuring technological patent scope by semantic analysis of patent claims – An indicator for valuating patents. *World Patent Information*, 58, 101906.

Wu, C.-C., and Leu, H.-J. (2014). Examining the trends of technological development in hydrogen energy using patent co-word map analysis. *International Journal of Hydrogen Energy*, 39(33), 19262–19269.

Yalçın, H. (2010). Bibliometric profile of journal of national folklore (2007–2009). *Milli Folklor*, (85), 205–211.

Yalçın, H., and Şeker, M. (2020). The effects of COVID-19 Pandemic on academic researches and publications. In Şeker, M., Özer, A., and Korkut, C. (eds), *Reflections on the Pandemic*. Ankara: Turkish Academy of Sciences, 217–240.

Yalçın, H., Shi, W. Y., and Rahman, Z. (2020). A review and scientometric analysis of supply chain management (SCM). *Operations and Supply Chain Management*, 13(2), 123–133.

Yalçın, H., and Yayla, K. (2016). Main dynamics of folklore discipline: A scientometric analysis. *Milli Folklor*, 2016(112), 42–60.

Yeh, H.-Y., Lo, C.-W., Chang, K.-S., and Chen, S.-H. (2018). Using hot patents to explore technological evolution: A case from the orthopaedic field. *Electronic Library*, 36(1), 159–171.

Yoon, B., and Park, Y. (2004). A text-mining-based patent network: Analytical tool for high-technology trend. *Journal of High Technology Management Research*, 15(1), 37–50.

Yu, W.-D., and Lo, S.-S. (2009). Patent analysis-based fuzzy inference system for technological strategy planning. *Automation in Construction*, 18(6), 770–776.

Zbikowska, H. M. (2003). Fish can be first – Advances in fish transgenesis for commercial applications. *Transgenic Research*, 12(4), 379–389.

Zeng, Y., Clerckx, B., and Zhang, R. (2017). Communications and signals design for wireless power transmission. *IEEE Transactions on Communications*, 65(5), 2264–2290.

Zhang, P., Yan, F., and Du, C. (2015). A comprehensive analysis of energy management strategies for hybrid electric vehicles based on bibliometrics. *Renewable and Sustainable Energy Reviews*, 48, 88–104.

Zhou, X., Huang, L., Zhang, Y., and Yu, M. (2019). A hybrid approach to detecting technological recombination based on text mining and patent network analysis. *Scientometrics*, 121(2), 699–737.

Zhou, X., Zhang, Y., Porter, A. L., Guo, Y., and Zhu, D. (2014). A patent analysis method to trace technology evolutionary pathways. *Scientometrics*, 100(3), 705–721.

Zikopoulos, P., and Eaton, C. (2011). *Understanding Big Data: Analytics for Enterprise Class Hadoop and Streaming Data*. New York: McGraw-Hill Osborne Media.

Zinda, K. (2004). *Patent data mining*.

Index